Ludwig Mauthner

The Sympathetic Diseases of the Eye

Ludwig Mauthner

The Sympathetic Diseases of the Eye

ISBN/EAN: 9783337036133

Printed in Europe, USA, Canada, Australia, Japan

Cover: Foto ©berggeist007 / pixelio.de

More available books at **www.hansebooks.com**

THE SYMPATHETIC DISEASES

OF THE

EYE

BY

LUDWIG MAUTHNER, M.D.
ROYAL PROFESSOR IN THE UNIVERSITY OF VIENNA

TRANSLATED FROM THE GERMAN

BY

WARREN WEBSTER, M.D.
SURGEON, UNITED STATES ARMY

AND

JAMES A. SPALDING, M.D.
MEMBER OF THE AMERICAN OPHTHALMOLOGICAL SOCIETY; OPHTHALMIC SURGEON TO THE MAINE GENERAL HOSPITAL.

NEW YORK
WILLIAM WOOD & COMPANY
1881

TRANSLATORS' PREFACE.

This comprehensive monograph, on the SYMPATHETIC DISEASES OF THE EYE, is the first of a series intended to embrace the whole province of Ophthalmology. The author, Dr. Ludwig Mauthner, of Vienna, a well-known specialist, has two objects in view: one, to compile, for the ophthalmic surgeon, the widely diverse opinions on the subjects under discussion; the other, to enable the general practitioner, and the student in ophthalmology, to gain an insight into the pathology, and especially into what should be the practical treatment, of the more important diseases of the eye.

Although the number of learned, conscientious, and skilled oculists in America is daily increasing, yet there will be exigencies in civil life, as well as in the military and naval service, when their assistance cannot be obtained. A large majority of patients affected with diseases or injuries of the eye, should, and naturally will, turn at once to their family physician for advice. The latter, with this monograph at hand, or others of the series as they shall appear, will be enabled immediately to judge of the triviality or of the serious nature of the case. He will then decide either to treat it himself, according to the latest light which scientific research and experience, as set forth in books, have thrown upon it, or to refer it, if haply he can, to a trustworthy specialist for more minute treatment, or for an operation, if necessary.

In so far as regards the subject of the present monograph (SYMPATHETIC DISEASES OF THE EYE), we may truly say that it is one of the most important with which the oculist is ever concerned. Upon his correct judgment will generally depend the future vision of the patient. Much more urgent, therefore, must be the necessity for general practitioners in the country, and for medical officers of the army and navy, to have at hand a clear and reliable description of the multiform symptoms, and the treatment, of Sympathetic Ophthalmia, so that they may at once recognize its presence, and treat it from the outset appropriately and effectually. Although cases of this nature are comparatively rare, their importance is sufficiently great to account for the appearance of this excellent work in an English version.

WARREN WEBSTER.
JAMES A. SPALDING.

PORTLAND, MAINE, September 1, 1881.

AUTHOR'S NOTE.

These " Lectures on Ophthalmology " cannot fully succeed in their professed object of popularizing, among practitioners of general medicine, the specialty to which the author belongs, unless he assumes that the readers have but slight acquaintance with ophthalmological terminology. He regrets, however, that he has occasionally been obliged to overstep the bounds of general description, and to adopt, for a time, the necessary minutiæ of his specialty.

MAUTHNER.

Vienna, March 27, 1878.

CONTENTS.

	PAGE
PRELIMINARY REMARKS, ...	9

SECTION I.
ANATOMY, ... 12

SECTION II.
ETIOLOGY, ... 17

SECTION III.
PATHOLOGY, ... 56

SECTION IV.
PATHOGENY, ... 105

SECTION V.
THERAPEUTICS, ... 146

INDEX, ... 209

THE
SYMPATHETIC DISEASES OF THE EYE.

It is a terrible thing when some constitutional disease, or a local disease outside the eye—perhaps of the brain—or some definite disease of the eye itself, or a traumatic agent, destroys the sight of both eyes at once. Then, again, it is lamentable when one eye is destroyed, at a greater or less interval after the other, from a repetition of the original injury, as has twice occurred in my experience, from the explosion of gunpowder, and the thrust of a cow's horn. The misfortune, however, is even more aggravated when the *second* eye is totally lost, simply from some disease or injury of the *first* eye; or when a surgical operation on the one eye not only fails of its object, but subjects the opposite eye to serious mischief; or when, after a successful operation on one eye, we attempt at a later date to gain some vision for the other, and not only find that the second eye is unimproved by the attempt, but also that, as a direct consequence of the last operation, the sight once happily restored to the first eye is again imperilled.

"Sympathetic ophthalmia" is a general term, which serves to designate, not a particular affection, but a whole series of ocular lesions, which differ from one another in their seat and manifestations, but always have a common origin. When an eye is laboring under injury or disease, it frequently happens that the other eye, which has hitherto been healthy, becomes, after a certain time, and without apparent cause, the seat of various functional or structural disturbances. The latter are called *sympathetic affections*, and, taken together, constitute *sympathetic ophthalmia*. Those diseases, therefore, which are superinduced in the *second* eye, upon an injury, or a disease, of the *first* eye, and which can be traced to *no other cause than the original injury or disease*, are regarded as *sympathetic diseases*.

Hardly any other province of ophthalmology is of more practical importance, and in no other are greater demands made, as well on the personal experience of the practitioner, as on his acquaintance with the experience of others; in hardly a second field is greater good to be expected from treatment, or greater evil from neglect, than in the one comprising the sympathetic diseases of the eye. Here it is not the fate of a single eye that is at stake, but the question that almost always confronts us is: Shall the individual suffer utter loss of sight, or shall the vision of at least one eye be wholly, or in part, preserved?

Before describing the symptoms of sympathetic affections of the eye, and their treatment, we must notice the primary injuries and diseases of the eye which most commonly excite sympathetic disturbances. First, however, it will be well to refresh, in a brief manner, our remembrance of the anatomical structure of the eyeball.

SECTION I.

ANATOMY.

THE *eyeball* is composed of several investing *tunics*, as well as of fluid and solid contents, called the *refracting media*. The most important of the latter is the crystalline *lens*, which is a double convex body, situated immediately behind the pupil, and having its axis in the same line with that of the eyeball itself. It is retained in its position chiefly by the *suspensory ligament* (*zonula Zinnii*), which connects its periphery with the anterior margin of the retina. The suspensory ligament is also attached to the ciliary body by a series of radiating folds or plaitings, into which the corresponding ciliary processes are received.

The *vitreous humor*, which occupies about four-fifths of the eyeball posteriorly, is surrounded by the retina as far forward as the termination of the latter, at the ora serrata, and is bounded, in front, by the ciliary body, the zonula of Zinn, and the *posterior capsule* of the lens.

That portion of the cavity of the eyeball which lies in front of the lens, between the latter and the cornea, is occupied by the *aqueous humor*. This space is divided into the anterior and posterior chambers by the *iris*, a thin, membranous curtain, hanging vertically in front of the lens, and perforated by the *pupil* for the transmission of light. The iris may be regarded as a process of the choroid, with which it is continuous, although there are differences of structure in the two membranes. The *anterior chamber* is bounded in front by the *cornea*, a perfectly transparent tissue, the innermost layer of which is a single stratum of flat, epithelial cells, which rest on the *membrane of Descemet*, and are bathed by the aqueous humor. The anterior chamber is bounded posteriorly by the ciliary ligament and the iris, and by that portion of the *anterior capsule* of the lens which lies free in the pupil.

At the place where the periphery of the cornea is overlaid, like a watch-glass, by the free edge of the *sclerotica*, a multitude of stiff fibrillæ stretch across, in a curved direction, from the inner surface of the cornea to the front of the iris, and constitute collectively the *ligamentum pectinatum iridis*. The epithelial cells covering the membrane of Descemet are continued upon the ligamentum pectinatum, where they form, in conjunction with the fibrillæ of the latter, a *cellular plate*, which separates the anterior chamber from the ciliary body.

The whole posterior surface of the iris does not lie directly in contact with the anterior capsule of the lens, but only the central portion, that is to say, the pupillary border. Hence, as the iris occupies a nearly level plane, its periphery is separated from the anterior convex surface of the lens, and the space known as the *posterior chamber* is formed. The individual *ciliary processes* project into the angle of the posterior chamber, in the region of the sclerotica. The posterior chamber is bounded in front by the iris, with its thick covering of pigment; whilst its posterior wall is made up of the anterior capsule of the lens, the zonula of Zinn, and the ciliary processes.

Inasmuch as the *pupillary margin* of the iris, in a healthy eye, moves freely over the anterior capsule of the lens, no obstacle exists to an interchange of the fluid contents of the anterior and posterior chambers; indeed, if the pupil be dilated by the instillation of atropia, so that the border of the pupil can no longer touch the anterior capsule, the two chambers become practically blended into one.

The *retina* is a delicate, semi-transparent expansion of the optic nerve, and extends nearly as far forward as the ciliary muscle, where it terminates by a jagged margin, the *ora serrata*. Its outer surface lies in contact with the pigmentary layer of the choroid; its inner surface, with the vitreous body. The *optic nerve* pierces the sclerotic and choroid coats at the back part of the eyeball, and enters its cavity at a spot called the *optic papilla*, a little to the nasal side of its posterior pole. On examining the concave inner surface of the retina, we observe, directly in a line with the axis of the globe, and situated about three millimetres outward from the optic papilla, a circular *yellow spot*, which presents a central depression (*fovea centralis*), in which the sense of vision attains its greatest perfection. A horizontal section of an eyeball, accurately dividing the optic papilla into an upper and a lower half, would not bisect the fovea centralis, which lies in a plane slightly below the papilla.

The *choroid* is the vascular membrane of the eye, and, with the ciliary body and iris, constitutes the *uveal tract*. It is interposed between the sclerotica and the retina, and is thinner than either of these tunics; but its important appendage, the *ciliary body*, which lies next to it in front, attains a considerable size, being about four millimetres thick from before backward. This body—which is made up of the zonula of Zinn, the ciliary processes, and the *ciliary muscle*—is divisible into two parts: the inner portion consists of the zonula and the ciliary processes; the outer portion (which was formerly regarded as a ligament, but in which the existence of muscular fibres has been demonstrated by Brücke, Bowman, and Müller) occupies the space between the scleral insertion of the cornea and the periphery of the iris. The ciliary muscle is united externally with the cornea and sclerotica, and, internally, merges into the ciliary processes; behind, it is continuous with the choroid, and, in front, is inserted, by a portion of its fibres, into the iris, whilst by others it is attached to the wall of the canal of Schlemm and to the ligamentum pectinatum iridis. The contraction of the ciliary muscle draws the choroid forward and (by aid of its circular fibres) inward, toward the equator of the lens.

During youthful life, or so long as the lens remains soft, its form is regulated by the degree of tension maintained in its capsule by the suspensory ligament. When the latter is relaxed, by the action of the ciliary muscle, the lens retracts by its own elasticity, and becomes more globular in shape, thereby increasing the refractive power of the dioptric apparatus of the eye. In a word, it is the office of the ciliary muscle to effect that adjustment of the eye (accommodation) for near and remote objects, which enables it to produce distinct images on the retina.

If we pass a probe from the outermost edge of the anterior chamber, through the ligamentum pectinatum iridis, into the ciliary body, we penetrate, beneath the cellular plate, a coarse-meshed net-work, lined with cells, analogous to the *canal of Fontana*, as found in the ox. This structure is to be distinguished from a circular canal, filled with venous blood, and called the *canal of Schlemm*, which is tunnelled out of the scleral tissue, around the margin of the cornea, and resembles, in places, a plexus of veins.

SECTION II.

ETIOLOGY.

THE ciliary body is copiously supplied with nerves and vessels, and may be called the dangerous region of the eye—the one from which most of the sympathetic affections of the second eye proceed.

The *diseases of the ciliary body* may arise either spontaneously or from traumatic causes. The association of a wound with the morbid process does not necessarily expose the second eye to increased danger. Nevertheless, a graver danger has been attached to the traumatic affections of the ciliary body, not only because they are more frequent than the idiopathic, but from the fact that when a foreign body remains in the eye the traumatic affections are less easily controlled, or, when apparently under control, are more readily rekindled. Wounds of the ciliary body should, indeed, excite solicitude, for they may, at longer or shorter intervals, inflict on both eyes the most unfortunate consequences. On the other hand, very serious accidents to the ciliary body have, under surgical treatment, or through some lucky and unforeseen accident, or even spontaneously, terminated in the recovery of the injured eye, without the implication of its fellow.

A patient came to me complaining that he had injured himself at smith-work, and that a piece of iron had certainly entered his eye. A small wound was visible in the upper and outer part of the sclerotica, near the margin of the cornea. The eye wept, showed slight episcleral injection around the cornea, and was sensitive to pressure at the wounded spot. A more careful examination showed that the lens was apparently clear and uninjured; no deeper wound nor perforation of the anterior chamber could be discovered. It was possible, however, that a small foreign body had penetrated the eyeball and still remained at the bottom of the wound. Perhaps it was lodged in the ciliary body, and, in that case, the inflammation excited therein (cyclitis) might endanger both the injured and the sound eye. A fine bistouri, introduced into the wound, under anæsthesia, encountered some metallic body. The wound was at once enlarged, and a small chip of iron removed with delicate forceps. All the signs of irritation disappeared with exceeding rapidity, the wound healed in a few days, and no sensitiveness whatever of the ciliary body remained.

In a second case, the patient had severely wounded his right eye while discharging a musket. He averred, with the utmost confidence, that no foreign body was lodged in the eye. But it was evident that a perforation, located in the centre of the cornea, had been made by a bit of an exploded percussion-cap. Had the fragment rebounded from the capsule of the lens, or had it, perchance, penetrated the lens itself? These points could not be then determined, for a large amount of pus occupied the anterior chamber and concealed the pupil. The iris was prolapsed into a puncture, which had been made in the lower border of the cornea for the purpose of evacuating the pus. It was in this condition that I first saw the patient. It was impossible, at that time, to decide whether the purulent masses which still occupied the pupil were nodules of exudation upon the anterior capsule, or were swollen and suppurating fragments of the wounded lens; the latter condition, however, seemed the more probable. Nevertheless, the pus gradually disappeared, and although the pupillary border of the iris was found extensively adherent to the anterior capsule, neither the latter nor the lens had been wounded. The eye continued to improve, but, along with some lachrymation and pain, a slight subconjunctival injection persisted around the dark-colored spot where the iris had prolapsed. One day, while examining the eye more carefully, in order to discover the cause of the obstinate irritation, I noticed that the dark prolapsed iris had a distinct metallic lustre, so that I at once suspected the presence of a piece of metal. With a pair of fine forceps I extracted, from a small excavation in the corneal edge of the sclerotica, where it lay imbedded, a rolled up piece of copper cap, 4 mm. long and $2\frac{1}{2}$ mm. wide. All the signs of irritation now disappeared in a very short time. A fortunate accident had saved both the wounded eye and its mate. The piece of metal had penetrated the cornea, struck the anterior capsule of the lens without opening it, and had then rebounded to the bottom of the *posterior* chamber, where it lay directly upon the ciliary body and excited a severe inflammation of the whole anterior part of the eyeball. The puncture of the cornea, which had been made for the removal of the pus from the anterior chamber, having luckily been *unskilfully performed*, a portion of the iris fell through the incision, and into the *pocket-like duplicature thus made the piece of metal was received.* After necrosis of the prolapsed iris the metal lay freely exposed at the edge of the cornea. Had the operation been made according to rule the iris would not have prolapsed, and the foreign body left within the globe would, in all probability, have produced a dangerous cyclitis, with the chance of involving the second eye.

The good results attained in the two injuries just described were due to surgical interference: in the one case, intentional, and, in the other, accidental. But sometimes severe wounds of the eye may terminate favorably, without any surgical interference whatever. A boy, twelve years old, was shot in the left eye with an arrow from the cross-bow of a playmate. The arrow stuck fast in the eye until pulled out by his companion. The eye reddened, but was not painful at first, and, immediately after the accident, the boy said that his sight was as good as ever. Four days later, on awaking from sleep, he noticed that he could see very little with the wounded eye, and, later in the same day, pain supervened, with almost complete blindness of the eye. On the next day the eye was examined by a surgeon, who found a small, round wound in the sclerotica, behind the lower and inner edge of the cornea. There was also pericorneal injection; the pupil was contracted; the unwounded lens was in its proper position; but the vitreous humor was clouded throughout. The tension* of the eyeball was normal, and no spot manifested any sensitiveness to the touch; but the vision was so reduced that light and darkness could barely be distinguished. The inflammatory symptoms soon became more marked, and pus, which must have come from the ciliary body, inasmuch as both cornea and iris were uninflamed, appeared in the

---

* By the word *tension*, which will be of frequent recurrence in these pages, we mean the feeling of hardness or softness of the eyeball, when we press upon it through the closed lids with the fingers. If the eye feels softer than the normal organ, we say the tension is diminished; if harder, the tension is increased.—TRS.

anterior chamber. Gradually, however, all the inflammatory symptoms subsided, and the turbid vitreous again became clear. Two years later, when I saw the boy for the last time, the ophthalmoscope revealed a very striking condition of things in the fundus of the eye. The retina, as well as all the rest of the interior of the eye, was visible, although somewhat indistinct. A large, dark cord extended from the optic papilla, directly through the vitreous body, to the point where the arrow had entered the eye. Immediately before its termination at this point, the cord divided into numerous slender threads. Its direction indicated the exact course of the arrow-head, which had, therefore, traversed the whole vitreous humor and become fixed in the optic papilla. A vascular neoplasm, which projected toward the vitreous, from near the insertion of the cord, appeared to have been due to the irritation in the papilla by the foreign body. The eye was free from any symptoms of irritation, and showed two-sevenths of the normal amount of vision, with a perfectly clear visual field.

We now have to note another important point. A foreign body imprisoned in the eye may prove a source of constant irritation for years, exciting from time to time severe inflammation of the wounded eye, and justifying the fear that sympathetic disease may at any time break out in the sound eye. If, however, during a violent attack of inflammation, the eyeball should unexpectedly open at some point, and the foreign body, so long present, be expelled from the eye, either spontaneously or by surgical assistance, a new and happy turn may be given to the case, affording permanent rest to the injured eye, and assuring the other from threatened destruction. I have, however, seen this favorable result but twice, the offending body, in each instance, being a fragment of glass.

In one case, a piece of glass—so large as to excite wonder that it could have either entered or occupied the cavity of the eyeball—came to light, after a violent inflammatory attack, and was finally extracted through the sclerotica, after the spontaneous opening had been greatly enlarged.

The second case was that of a woman who applied for an artificial eye. A splinter of glass had flown into her left eye, in early youth, and had ever since been a source of constant irritation, provoking severe inflammatory attacks in the affected eye, and greatly impairing the vision of the opposite eye. She reported that, during a violent inflammatory exacerbation, the splinter had appeared at the surface and been spontaneously expelled. After that event the injured eye gave no further trouble, and the second eye could be used for all sorts of work.

The *injuries of the ciliary body* and its vicinity, known to give rise to *sympathetic* disease, are appropriately classified as *accidental* and *operative* injuries.

The *first* of these divisions comprises: penetration of foreign bodies into the ciliary body, with lodgement therein; punctured or incised wounds of the ciliary body, without lodgement of a foreign body; contused or lacerated wounds of the ciliary body, inflicted by blunt agents; incised, punctured, and lacerated wounds of the periphery of the cornea, with or without injury of the ciliary body, whereby the periphery of the iris alone, or along with it a portion of the ciliary body, becomes incarcerated in the wound; and finally, contusions of the ciliary body, from mechanical violence applied to the eyeball, without opening it.

A foreign object lodged in the ciliary body may sometimes become encapsuled, and so be made innocuous. When this happens, the diagnosis of its presence is certainly very difficult. Bowen, however, has observed (1875) that such an object, after a long and harmless stay, may suddenly and dangerously announce its presence. A particle of iron, the size of a pin-head, lay among the fibres of the ciliary muscle for *nine* years, causing extensive thickening in its neighborhood, as was found at a subsequent examination. After this long period, pain in the ciliary body was felt, on pressing the spot where the injury had been inflicted, and, a few weeks later, sympathetic ophthalmia supervened, which only ceased after enucleation of the wounded eye.

It thus appears certain that a foreign body, either

free or encapsuled, may harmlessly remain for a long time, and even for life, not only in the ciliary body, but in any other part of the eye. It is, on the other hand, no less clear, from another case, also reported by Bowen (1875), that the wounded eye, even after a very protracted interval of quiescence, gains no certain immunity from severe inflammation and ensuing sympathetic disturbance, liable, as they both are, to be caused by the presence of the original foreign body. In the latter case, a piece of metal, two and a half millimetres long, lay imbedded in the optic nerve for *seventeen* years, and it was only after it had produced inflammation and disorganization of the uveal tract, that sympathetic phenomena—intolerance of light, ciliary injection, and discoloration of the iris—appeared in the uninjured eye.

Although the injuries of the ciliary body are much more dangerous than analogous injuries of other parts of the eye, from the greater proneness of the former to develop the severe train of symptoms presently to be described, yet a simple injury of the ciliary body, *when not complicated by prolapse of the iris, or incarceration of some portion of the ciliary body* in a penetrating wound, is not often followed by serious consequences.

Violent contusions and concussions, inflicted upon the eye by blunt bodies—for example, the naked fist, or one in which the fingers are covered with heavy rings—play relatively the *most frequent* part in the etiology of cyclitis, and its associate diseases, irido-cyclitis and irido-cyclo-choroiditis. *Next in order of frequency* come penetrating and cutting wounds, without prolapse or constriction of any of the parts; and *least frequently* of all (whatever may be their inherent danger), the penetration and permanent location of small foreign bodies within the ciliary body.

The symptoms and anatomical changes set on foot by injuries of the ciliary body may be so insidious and painless, at the start, as to be quite unrecognizable. Soon, however, more marked symptoms appear: the injured eye becomes intolerant of light and bathed in tears, while a ring of blood-vessels environs the cornea. If we touch the ciliary region with a blunt probe, or simply press with the finger through the closed lids, the patient complains of its sensitiveness, and, in particular spots, of acute pain. The cornea becomes hazy and dull on its external surface, and the iris, if visible through the cornea, is seen to be discolored, its natural lustre gone, and its striated appearance obscured. The pupil is still open, but atropia no longer exerts any influence upon its size. We soon discover foundation for our suspicion that the pupillary edge of the iris is adherent to the capsule of the lens; while the whole posterior chamber is filled with inflammatory exudation, gluing the iris, the ciliary body, and the anterior capsule firmly together. Pus may occupy the floor of the anterior chamber, having forced its way directly thither from the ciliary body, through the ligamentum pectinatum iridis and its cellular plate. If the pupil be still sufficiently clear to permit of the use of the ophthalmoscope, we can with difficulty distinguish the fundus of the eye through the intervening turbidness. So long as this opacity is still *diffuse*, it is hard for the observer to decide how much of it depends on the cornea, as well as on the turbid aqueous full of pus-corpuscles, or how much on the vitreous. But when dark objects, of varying size and shape, float about in the affected eye upon its being quickly moved to and fro, we know that the vitreous humor is involved in the pathological process. Vision, meanwhile, has diminished exceedingly.

The eyeball now becomes ominously soft to the touch, and the acuteness of vision markedly diminished. The anterior chamber is narrowed, inasmuch as the lens is pushed forward toward the already turbid and flattened cornea. The periphery of the chamber may, however, appear deeper at places than normal, inasmuch as the masses of exudation which occupy the posterior chamber have formed a cicatricial tissue between the iris and anterior capsule, become consolidated, and so dragged the ciliary border of the iris backward toward the lens. The iris itself, having passed through its stage of proliferation and softening, is now atrophied, and turned to a dirty, yellow color. The black pigment which lines its posterior surface is visible through the anterior layer, giving it a dotted appearance, while here and there tortuous veins are displayed, owing to the inflammatory swelling of the ciliary body, whereby the venous blood of the iris is now impeded in its passage to the choroid. The pupil may, at this stage, be still permeable for light, but more frequently it is blocked with masses of exudation.

The morbid process culminates when the inflammation of the ciliary body (cyclitis) is communicated backward to the choroid (choroiditis), which, in turn, involves the contiguous retina (retinitis), whilst the nutrition of the deeper structures of the eye becomes so disturbed that a marked reduction in the mass of the vitreous humor takes place. The direct consequence of the atrophy of the vitreous is the loss of the normal tension of the globe, which now feels soft, and may become so flaccid as to be indented at the places corresponding to the recti muscles. But even after phthisis of the entire eyeball, with total inflammatory destruction, or even detachment, of the retina, and consequent extinction of vision, the eye does not subside into quiescence. The offending ciliary region may still be tender and irritable to the touch, painful upon the slightest occasion, and a source of constantly impending danger to the other eye.

We have, moreover, to mention Mooren's assertion

that after the introductory symptoms, such as pericorneal injection, photophobia, lachrymation, and partial sensitiveness of the ciliary body, in a *typical case of simple acute cyclitis*, we *first* see an *increase of depth in the anterior chamber*, due to the inflammatory adhesion of the periphery of the iris to the ciliary body. We are also struck by the fact that *no iritic adhesions* to the anterior capsule, even *at the pupillary border*, exist at this time, the *pupil being readily dilatable* by the instillation of atropia. Should the retraction of the periphery of the iris progress, then the veins of the iris dilate, the aqueous humor becomes cloudy, pus appears in the anterior chamber, and opacities quickly and copiously form in the vitreous humor.

When, in connection with an injury of the ciliary body, the eyeball is opened by a punctured or incised wound, or is lacerated and contused by some blunt instrument (cow's horn), or a projectile, the injury is usually complicated by a prolapse, into the wound, of a portion of the ciliary body, or the periphery of the iris, or both together. In the majority of such cases, the cyclitis, or irido-cyclitis, is directly produced by the injury, and not by the incarceration of the ciliary body or iris. Wounds of this kind are sometimes very remarkable. I once saw an eye that had been bitten by a horse, so that the organ was lost, after violent symptoms of cyclitis, and the other eye subsequently suffered from severe sympathetic ophthalmia. Lebrun

(1870) reported a case in which a leech, applied to the neighborhood of an eye for therapeutical purposes, strayed to the edge of the cornea, where it inflicted a bite that was followed by sympathetic symptoms in the other eye. We have already mentioned (page 20) an extraordinary case in which a foreign body flew through the cornea, as far backward as the anterior capsule, from which it fell to the floor of the posterior chamber, and there rested in menacing contact with the ciliary body.

Both contusions and perforations of the eyeball may cause *cyclitis in an indirect way.* Thus, a contusion may partially lacerate the suspensory ligament (zonula Zinnii), so that the lens may either sink downward upon the ciliary body, and excite irritation by its contact with the latter, or it may drag upon the ciliary body through its remaining attachments to the zonula, and produce a similar effect. Again, when a foreign body has penetrated the lens, or extensively lacerated its capsule, the fragments of the mutilated lens may fall into the bottom of the *posterior* chamber, and cause severe inflammation of the iris and ciliary body. If, however, the fragments of the lens fall into the *anterior* chamber, their presence usually provokes much less inflammation. Thus may injuries of the eye lead indirectly, through lesions of the lenticular apparatus, to disease of the uveal tract, and, later, to sympathetic affections of the opposite eye.

We must here remind ourselves that it is not only the accidental injuries of the eye, but also those which are incidental to *surgical operations*, that *may initiate sympathetic ophthalmia*. Among the operative injuries, the one called *iridodesis*, and the various *operations for cataract*, occupy the first rank. Critchett (1858) devised the operation of iridodesis, with a view to provide the disabled eye, under certain circumstances, with better vision than could be gained by iridectomy.

The operation called *iridectomy* consists in making a new opening in the iris for the rays of light to enter the eye, when the natural pupil is covered by a central opacity of the cornea, or when the pupil lies in front of a stationary central cataract. A piece of the iris is excised, so that a portion of the still transparent cornea, or lens, faces the artificial opening. This operation, when performed for optical purposes only, has not usually given satisfactory results. It is, indeed, invaluable when the central opacity of the cornea wholly conceals the pupil, and is at the same time completely or nearly opaque, provided that the outer portion of the cornea, which *appears* normal, is *really* so, as regards both transparency and curvature. Moreover, in the rare disease called stationary nuclear cataract, in which the central portion of the lens lying directly behind the pupil is totally opaque, and a considerable margin of the lens beyond the opacity is perfectly transparent, iridectomy is a reliable resource. But such clear indications for the operation are seldom met with, for the offending spot in the centre of the cornea oftentimes falls far short of complete opacity, whilst the central cataract, on account of which the patient demands "more light," is almost always of the so-called lamellar variety, in which an opaque lamella or zone intervenes between the nucleus and cortical portion, which are both clear. In many cases the impairment of vision is so slight as not at all to interfere with ordinary pursuits, and no surgical operation is warrantable under such circumstances. Furthermore, the lamellar variety of cataract, even in its extreme degree of development, still permits a certain amount of light to enter the interior of the eye. If, therefore, an iridectomy is performed on an eye affected with an incomplete opacity of the cornea or lens, the retina receives light not only through the newly made aperture, but through the old pupil. The failure of the opacity to prevent the transmission of light through the original pupil is a source of disturbance to the eye as an optical apparatus, because in the eye, as in the camera obscura, clearly defined images are only produced when all irregularly refracted rays are excluded. When diffused light is thrown over the retinal image, the latter becomes indistinct. For the foregoing reasons, the performance of iridectomy, under the circumstances above mentioned, does not enable the eye to see well; for not only does diffused light continue to reach the retina, but the dazzling sensation caused by too brilliant illumination of the field of vision is aggravated by the operation, inasmuch as the pupil is thereby not only deprived of its contractile power, but its area is greatly enlarged.

On account of the excessive size of the pupil, its loss of reactionary power, and the disturbance of the retinal image by diffused light, which follow iridectomy, an attempt was made to obviate these evils by substituting the operation called *iridodesis*, in performing which a small incision is made in the cornea, close to the sclerotica, and the peripheral portion of the iris seized and drawn out of the wound, with such precautions that the entire pupillary border is left within the anterior chamber. A loop of thread is afterward tied around the prolapsed iris, to prevent it from slipping back into the eye; the strangulated piece then rapidly necroses, falls off with the thread, and the wound is soon healed. The pupil has thus been transformed into an oval or longitudinal opening, and moved *in toto* toward the place of incision, the portion of the iris directly opposite the place of incision having been stretched to permit of the dislocation. The displaced pupil, with its constrictor pupillæ intact and its reactionary function unimpaired, covers a scarcely greater area than it did before the operation. Moreover, a portion of the iris is interposed, as a diaphragm, be-

hind the semi-opaque corneal spot, or *in front* of the partly translucent cataract, thereby protecting the retina from diffused rays of light; so that, barring the sacrifice of some trueness of the corneal curvature (an evil which Pagenstecher sought to avoid by removing the incision into the verge of the sclerotica), we now have an eye which, although not projecting an absolutely perfect image upon the retina, certainly possesses better vision than it would have, had an iridectomy been performed.

Wecker practises iridodesis in those cases in which the lens, from whatever cause, has become dislocated, so that its centre no longer corresponds to the centre of the pupil, or, more strictly speaking, to the axis of vision. If, for example, the zonula has been torn at its lower and inner insertion, the lens is displaced upward and outward, so that the space thus left between the lower internal border of the lens and the adjacent ciliary processes is partly visible through the pupil when dilated with atropia, or even when of normal size. Two images of an object, seen with such an eye, are thus projected upon the retina: one of them by the cornea, aqueous humor, *lens*, and vitreous humor; and the other, by a refractive system from which the *lens* is absent. If the image made without the aid of the lens be, for any reason, the more useful of the two, the operation of iridodesis enables us to transfer the pupil permanently to a part of the cornea behind which the lens is absent, whilst, by the same procedure, the iris on the opposite side is stretched over the dislocated lens, so as to cut off the second image, which would otherwise interfere with distinct vision.

Iridodesis was at first regarded as a perfectly safe operation. But, in 1863, Alfred Graefe published the following significant case, in which iridodesis was performed on the eyes of a workman, aged twenty-three. Both eyes of the patient were affected with lamellar cataract, which, however, still permitted him to read No. 3 of Jaeger's test-types. Vision was improved immediately after the operation; but eight weeks later Graefe found the man *blind in both eyes,* with occlusion of the pupils, in consequence of irido-cyclitis. The eyes, however, were not soft. The patient could see well with both eyes during the first week, at the end of which time, without any apparent cause, the sight diminished, first in one eye, and very soon thereafter in the other, until it was reduced, at the time of the examination, to a merely quantitative perception of light. The exciting cause of the irido-cyclitis, in each eye, was attributed by Graefe to the stretching of the iris, incident to the iridodesis. Did not, however, *sympathetic* inflammation play its rôle in this case? It is possible that the operation had *directly* excited iridocyclitis in *one* eye only, and the inflammation had extended *sympathetically* to the other, so that the same lesions would have appeared in the second eye even if it had not undergone the operation. Although the nearly simultaneous involvement of the two eyes, in Graefe's case, makes the latter opinion less trustworthy, nevertheless, as we know from experience that irido-cyclitis is prone to be followed by sympathetic disease, and as Graefe established the existence of irido-cyclitis depending on the iridodesis, it must be admitted that this operation is not devoid of both primary and sympathetic danger. In fact, soon after Graefe's case came to light, one was reported by Steffan (1864), in which a girl aged nineteen, who had undergone iridodesis in one eye only, was affected, five weeks after the operation, with irido-cyclitis in both eyes. The disease may have first appeared in the wounded eye so insidiously as to receive no attention; but it was not until the affection had, some weeks later, extended to the hitherto perfectly sound eye, that the patient applied for relief.

When, during a visit to London in 1864, I mentioned to Critchett that the unfortunate cases of Graefe and Steffan had produced a want of confidence in iridodesis, among German oculists, he was not a little surprised at the two failures, as he had never encountered like results in his very large personal experience with the operation. My own operations have, likewise, been successful. But, on the other hand, unsuccessful cases and unfavorable criticisms of iridodesis have been sufficiently frequent in ophthalmological literature, since 1864, to place the operation where it now remains—in discredit.

Of far greater importance than iridodesis, as regards the danger of exciting inflammation, which may be propagated sympathetically to the second eye, are the *operations for the relief of cataract.*

One of the fundamental methods of operating for this disease, that of *depression* or reclination, by which a hard cataract is forcibly thrust away from the axis of the visual rays into the vitreous body, is now almost totally abandoned on account of the destructive consequences that ensue, not only in the operated eye, but, secondarily, in its fellow. The displaced lens often plays the part of a foreign body—resting, it may be, in disagreeable contact with the ciliary body and choroid. It may thus lead to inflammation of the uveal tract, if, indeed, this condition has not already been set up by the operation itself. The bad repute into which reclination has fallen is, however, due rather to the danger incurred by the eye undergoing operation than to an appreciation of the sympathetic disturbances that may subsequently develop.

Nor are the two operations of *division* and *extraction,* so extensively employed in our days, wholly devoid of analogous risks. The object of division or discission of a cataract is to lacerate the anterior capsule and break up the substance of the lens, so that the latter shall come into contact with the aqueous humor.

If the lens be only partially opaque, as in lamellar cataract, it becomes wholly so soon after exposure to the aqueous, and its fragments are gradually dissolved and absorbed until the cataract disappears. It sometimes happens, either when proper precautions have not been taken during the operation, or in spite of them, that the lenticular fragments imbibe a great deal of aqueous humor, swell considerably, press upon the iris, and cause severe iritis, followed rapidly by cyclitis and possibly by sympathetic disturbances. Although *division* is regarded by oculists as a very imperfect surgical procedure, there are, nevertheless, a few forms of cataract to which no other is so well adapted. Among these are the lamellar cataract and the extremely rare variety called stationary nuclear cataract, in both of which the transparent periphery of the lens adheres so intimately to the capsule that it cannot be removed by the extraction method, with satisfactory results.

Extraction by the *flap operation*, and v. Graefe's method of *modified linear extraction*, are the two most important of the different surgical operations for the removal of cataract. In operating by the first-named method, a semicircular flap, involving the whole upper half of the cornea, is made by incising the latter close to its scleral border. A large, patulous wound is thus produced, through which the lens is evacuated by gentle compression of the globe. In v. Graefe's

ETIOLOGY. 39

method of modified linear extraction the peculiarity of the incision is that it lies entirely in the sclerotica, and does not form a flap, its only curve being that of the eyeball itself. The incision is from ten to twelve millimetres long—its middle point lying at the topmost point of the corneal margin. Through this incision the lens is removed, after a preliminary iridectomy and laceration of the anterior capsule. The operations now most in vogue are a sort of compromise between the old flap operation, and the genuine peripheral linear extraction as modified by v. Graefe.

When the old method of removing the opaque lens by the flap operation was generally practised, very little was said of sympathetic ophthalmia after operations for cataract. Now and then we heard of irido-cyclitis and sympathetic affections, after the operation, and, in fact, a few such cases are matters of record; but we undoubtedly hear much more of sympathetic disturbances in connection with operations for cataract, since the era of linear extraction.

In all probability the first enucleation of an eyeball, upon which the method of linear extraction had been practised, was one that I performed in 1867, on account of sympathetic ophthalmia of the other eye. A cataractous lens had been removed with complete success, by the flap operation, in 1865, from the left eye of a man fifty years old. One year later Jaeger operated on the right eye by a linear method (the curved-lance section). Although the operation was skilfully performed, without any prolapse of the vitreous humor, irido-cyclitis set in, and was followed by atrophy of the eyeball. Thirteen months after the second operation the patient again applied for relief, the atrophic eye having never become quiescent, and being still affected with pain and photopsies. Six weeks prior to his reappearance pain commenced in the *left* temple, and, later, invaded the whole side of the head, undergoing exacerbations and remissions, but never complete suspension. Along with these symptoms, the vision of the left eye (which, as the patient declared, had been better, with the aid of cataract-glasses, since the first operation, than ever before) became impaired, and, at the date of examination, was reduced to one-fourteenth of normal, whilst the whole field of vision was obscured by a thick mist. The tension of the left eye was natural; both cornea and iris were of healthy appearance; but the vitreous humor, when illuminated by the ophthalmoscope, was seen to be turbid throughout. After enucleation of the right eyeball, the sympathetic symptoms and the ophthalmoscopic appearances gradually improved; but no amendment of vision had taken place at the time of the patient's discharge, nine weeks after the operation. In the enucleated eye the anterior portion of the choroid, with the neighboring part of the ciliary muscle, could be easily detached from the sclerotica, whilst the connection between the retina and the vitreous body was likewise abnormal. In this unfortunate case the extraction of a cataract from the second eye had not only failed of its immediate object, but had seriously endangered the restored sight of the first eye.

Knapp reported a similar unfortunate case in 1869. He operated successfully, after v. Graefe's method, on the left eye of a man sixty years of age. The eye healed in a favorable manner, and, six days after the first extraction, the operation was repeated upon the other eye. The repetition, however, was less fortunate. There was hæmorrhage into the anterior chamber, with subsequent iritis, and, later, sympathetic iritis of the first eye. Six weeks after the operation, both pupils had become occluded, and both eyeballs somewhat soft.

When the subject of sympathetic ophthalmia, occurring after cataract operations, was introduced by Klein, at the Heidelberg Ophthalmological Congress in 1874, a whole series of cases, wherein sympathetic affections had proceeded from the linear, or the various modifications of the linear, extraction, were communicated by oculists present. Becker collected (1875) twenty-two cases (neglecting, however, to include Knapp's case) of sympathetic disease, resulting after cataract operations. Seven of these cases followed the flap operation, four of the latter being well-recognized specimens of simple senile cataract; and fifteen occurred after operations by the linear method. Since that time further reports have been made of cases of sympathetic disease resulting from v. Graefe's extraction method.

The various causes of the original irritation in an eye that has been subjected to an operation for cataract are: incarceration of the iris in the wound, with or without visible prolapse of the iris (Klein, v. Arlt); imprisonment in the wound of a portion of the capsule of the lens, so that the suspensory ligament and ciliary body, at the opposite side of the eye, are dragged upon, or detachment of the ciliary body at the same spot (Horner); shrivelling of the capsule of the lens (caused by inflammatory exudation, or the development of a secondary cataract), with subsequent stretching of the iris and ciliary body (Hänel, Becker); and lastly, direct injury of the ciliary body, when the incision has been made too far out in the sclerotica (Ed. Meyer).

Shall we include simple *iridectomy* among the surgical operations that may cause sympathetic ophthalmia? Individual cases, showing this origin, are on record.

We have, so far, seen how traumatic affections of the *uveal tract* may endanger the integrity of the uninjured eye, and it is now time for us to inquire what importance those *affections* of the same regions, which are *not due to injuries*, may have in the production of sympathetic phenomena. The affections not due to injury are divisible into two classes: the one embracing diseases excited by *mechanical irritation* of some portion *of the uveal tract* by bodies which cannot, strictly speaking, be designated as traumatic agents; and the other, comprising the *purely idiopathic affections*.

In the first class belong those lesions which are produced by spontaneous dislocations of the lens, as well as by cysts of the iris, choroidal sarcomata, retinal gliomata, and intra-ocular cysticerci. Mooren believes that irido-choroiditis is produced by a spontaneously dislocated lens, only when the latter has fallen into the *anterior* chamber. Hulke, Knapp, and Nagel saw cases in which cysts of the iris had caused irido-choroiditis, with sympathetic irritation; and moreover, implication of the second eye, even where the first eye never became inflamed. An eye affected with choroidal sarcoma is prone to be succeeded by sympathetic disease (Pagenstecher, Norris, Steffan, Nettleship, Salvioli, Hirschberg, Knies); but it should be borne in mind that choroidal sarcoma is very frequently due to a traumatic agency. Steinheim reports a case of sympathetic irido-choroiditis ensuing upon traumatic glioma of the retina. The cysticerci are analogous to the neoplasms, in their causal relations to primary irido-choroiditis and its sympathetic sequelæ.

Idiopathic cyclitis, or irido-cyclitis, is a rare disease. When its attacks upon the two eyes are not synchronous, but are separated by a certain interval, it is not always easy to determine whether the second eye is sympathetically affected, or the disease in both eyes is due to a common cause. The same may be said of irido-choroiditis occasioned by the *syphilitic* poison, inasmuch as the disease of the second eye may be a sympathetic, and not a syphilitic lesion. When attacks of that variety of irido-choroiditis, which sometimes attends *cerebro-spinal meningitis*, occur simultaneously in both eyes, the operation of a common cause is evident; but if, on the other hand, one eye is first destroyed, and, later in the disease of the nervous system, the other is attacked in a similar way, it is probable that sympathetic influences have been at work. Noyes has reported a curious case of *herpes zoster ophthalmicus* of the left eye (that variety of herpes zoster in which the eruption follows the distribution of the chief cutaneous branches of the trigeminal nerve), in consequence of which both eyes were destroyed by subsequent irido-choroiditis, beginning in the right eye ten months later than in the left, and without herpetic disease of the former. Jeffries, likewise, saw a case of *temporary sympathetic disturbance* transplanted from an eye that had been destroyed by the same variety of herpes zoster.

If prolapse, or incarceration of the iris or ciliary body, within a traumatic opening of the eyeball, near the corneal margin, may provoke irritation of the uveal tract and sympathetic phenomena, it is easy to understand how the same effect may be produced when one or other of these structures is prolapsed or incarcerated within a similarly situated opening, made in the eye by an *ulcerative process*. In the latter condition very much the same relation of parts exists as after iridodesis; and indeed, this sort of natural displacement of the pupil is quite frequent. But we must guard ourselves against an exaggerated conception of the danger involved in the accident. I cannot recollect a case, in my personal experience, in which I have seen serious results to the second eye ensue upon this kind of cicatrization of the iris, even of its peripheral portion, in the cornea.

Where the ciliary body is thus imprisoned, a much more encouraging prognosis can be made than after a traumatic injury, inasmuch as the latter very frequently superadds a direct wound of the ciliary body. The danger of sympathetic inflammation is further diminished when the ulcerated perforation of the cornea is *very large,* so that, instead of a small strangulation, a great part of the iris protrudes through the cornea, becomes indurated and thickened from exposure, and forms a permanent protuberance (staphyloma) throughout the area of the absent cornea. In the same way, severe chronic inflammatory processes in the eye may cause the sclerotic zone, just outside the margin of the cornea, to become relaxed and softened, so that the intra-ocular pressure pushes it forward in such a manner as to present a series of small staphylomata, surrounding a greater or less arc of the corneal periphery, and in some cases even its whole circumference. If, under such circumstances, sympathetic symptoms should declare themselves, they must be attributed, not so much to direct stretching and laceration of the ciliary body, as to a defect in the suspensory ligament, somewhere around the equator of the lens, permitting the dislocation of the latter, and the consequent development of sympathetic phenomena in the manner before described (page 30).

So far as we have at present proceeded, it has appeared that the inflammatory lesions of the *uveal tract* threaten most danger to the second eye. None of the forms of uveal inflammation here brought under notice have followed a turbulent course, nor have they been attended with any *acute purulent* process. Their character has been insidious, and the ciliary body has been always more or less directly involved.

Glaucoma, simply as such, possesses no inherent power to awaken sympathetic disease. When, however, in the last stages of glaucoma, cyclo-choroiditis sets in, and the eye, hitherto abnormally hard, becomes soft, as well as painful over the ciliary region, the second eye becomes as much endangered sympathetically (Mooren, v. Arlt) as if the cyclo-choroiditis had its seat in a non-glaucomatous eye. Moreover, when we see a case of sympathetic ophthalmia ascribed to a detachment of the retina in the first eye, or to a hæmorrhage into the vitreous humor, we should incline toward the belief that a cyclitis had supervened upon the primary lesion, and had itself been the cause of the sympathetic derangement; as Mooren expressly argues, in the case of retinal detachment.

Some important questions which we next have to answer are: Does *purulent* inflammation of the uveal tract, also, lead to sympathetic ophthalmia? Can sympathetic ophthalmia supervene when the disease of the first eye is confined to the iris or to the choroid alone, produces no tenderness over the ciliary region, does not implicate the ciliary body, and pursues an unobtrusive course? Finally, can sympathetic ophthalmia be set on foot without lesion of any part whatever of the uveal tract of the first eye?

It has been generally held that *acute purulent* inflammation of the uveal tract (better termed *panophthalmitis*, inasmuch as the purulent process, accompanied by great swelling, is not confined to the uveal tract, but attacks all the tunics of the eye, as well as the vitreous humor) is devoid of sympathetic danger to the second eye. But exceptional cases to the contrary have been reported (Mooren, Rossander). Alt, who ascertained the pathological histories of one hundred and ten eyes, which had been enucleated on account of sympathetic disease (thirty-two of them under his own observation), found that twenty-one of the numa ber, or nineteen per cent., had been affected with typical panophthalmitis.

Again, it has been established that sympathetic affections may occur independently of any disease of the ciliary body, and even without any well-defined lesion of the uveal tract. Mooren (1869) cites among the diseases which may lead to sympathetic trouble, not only lesions of the ciliary body, but also those of the conjunctiva, sclerotica, cornea, iris, choroid, retina, and lastly, atrophy of the globe. It should be added, however, that he gives the most etiological importance to cyclitis, and lays particular stress upon the stretching, or laceration of the ciliary body, whereby a simple, minute prolapse of the iris becomes fraught with danger to the second eye. Peppmüller (1871) reported a few cases of sympathetic iritis following simple prolapse of the iris, without symptoms of cyclitis. Lüders (1872) saw a case of injury of the eye, in which the iris and anterior capsule of the lens, in the second eye, became agglutinated together seven weeks after the injury, although there had been no sensitiveness to pressure or softening of the injured eye.

From a series of cases brought forward by Warlomont (1872), it appeared, in one case, that an obstinate sympathetic kerato-conjunctivitis could only be cured after enucleation of the first eye, which had been for a long time atrophied, but *never sensitive*. In another case, that of a veteran, Warlomont speaks of a " severe external inflammation of the right eye " as an expression of sympathetic disease, although the stump of the other eye, which had been destroyed by a wound, was "perfectly painless." Other cases of the series give abundant evidence that phthisical eyeballs, which have never manifested pain, either spontaneously or on pressure, can, nevertheless, set up sympathetic disease. Out of ninety cases of sympathetic ophthalmia, published by Rossander in 1876, two originated in painless atrophy of the fellow eye; and out of ninety similar cases, reported by Vignaux in 1877, eight could be clearly referred to the same condition. The statement, therefore, is not entirely warrantable, that when a phthisical eye has seemed perfectly quiescent, a deposit of bone within the degenerated globe, irritating the choroidal tract in a purely mechanical way, and thereby renewing the tenderness and pain in the atrophic eye, must *invariably* be present in order to produce sympathetic disease.

Cohn (1871) met with two cases of sympathetic impairment of vision, after *gunshot wounds*, without symptoms of iritis, or cyclitis, in the wounded eye. In one of his cases the blind and offending eye had undergone extensive inflammation of the choroid and retina, as was established both by the ophthalmoscope and by anatomical examination after enucleation. In the other case, only a superficial grazing wound from a fragment of shell, had been inflicted upon the eye, which showed no internal lesion other than an effusion of blood between the yellow spot of the retina and the choroid. Brecht (1874) also saw a case in which disturbance of sight in the right eye had been transmitted sympathetically from its injured fellow, which latter, however, was "absolutely quiescent, showed no trace of unnatural redness, and was wholly devoid of pain, either spontaneously or under pressure." Pflüger (1875) traced a sympathetic affection of the one eye to a wound made by a piece of stone on the other, whilst the ciliary body of the injured organ seemed to be normal in every respect. He also reported another case, at the same time, in which an eye that had been destroyed by gonorrhœal ophthalmia, proved treacherous to its mate a few weeks later; nevertheless, when enucleated, it showed no sign of cyclitis, but simply an inflammatory infiltration of the iris.

Indeed, if we give credence to general pathologico-anatomical reports, we shall not need to search out individual cases in order to prove that sympathetic affections of the eye may arise quite independently of *any* disease of the ciliary body. Out of one hundred and ten dissected eyes upon which Alt reported, in the "*Archiv für Augen- und Ohrenheilkunde,* 1877," only seventy-six and one-half per cent. disclosed any disease of the ciliary body. Alt's words are as follows: "The iris is altered in sixty-eight per cent., and the choroid in seventy-three per cent. of the cases; so that the alterations found in the individual parts of the uveal tract are about equally distributed—those in the ciliary body very slightly exceeding in number those in each of the other parts."

The fitting of an *artificial eye* upon a painless stump has been known to develop sympathetic ophthalmia (Lawson, Mooren, Keyser); and reports of cases are at hand where the insertion of an artificial eye into an orbit, from which a diseased eye had been removed to abolish sympathetic irritation, has again excited the same morbid condition (Salomon, Warlomont). Finally, it has happened that the *enucleation* of an injured eye, or the sequelæ of the operation, performed for the especial purpose of preventing the sympathetic implication of its partner, have *produced* the apprehended condition (Mooren, Colsmann); or that the amelioration first following the enucleation of the offending eye has afterward disappeared, and the sympathetic disturbance been reinstated by the agency of the surgical operation itself (Hasket Derby).

While it is already evident, from our superficial notice of facts, which will receive further consideration as we proceed, that manifold forms of *sympathetic disease may arise without the presence of cyclitis at the time, or even without disease of any portion of the uveal tract*, there remains a question which should be answered in this place. Assuming that an injured eye, in which no foreign body lies concealed, recovers *perfectly*, so far as we can ascertain by clinical examination, from an attack of severe cyclitis—recovers even without degenerating into a state of atrophy—can such an eye, nevertheless, excite symptoms of sympathetic ophthalmia in the fellow eye? This question, be it understood, can only be put where the cyclitis is of *traumatic origin;* for if, after the recovery of an eye from *spontaneous cyclitis*, the same disease be set up in the second eye, we cannot have absolute proof of its sympathetic character. In answer to the foregoing question, I communicate the following case:

A common laborer, sixty years of age, presented himself at the Ophthalmic Clinic, October 3, 1875. He stated that he had been struck on the right eye, five years previously, by the rebounding branch of a tree, and that the sight of the injured eye had been instantaneously lost. He also complained that for about five years preceding his appearance he had been unable to read with his left eye, and that during the last year the sight of this eye had rapidly decreased. Both eyes showed signs of cataract. In the totally opaque lens of the *right* eye sparkling crystals of cholesterine, the product of a prolonged process of degeneration, justified the inference that the cataract had existed even a considerable period previously to the infliction of the injury. In the left eye the cataract was of more recent formation. In each eye the perception of light corresponded to the degree of opacity of the lens. Both cataracts were extracted at one sitting, by v. Graefe's method, but both operations met with impediments to their perfect performance. In the right eye fragments of the lens remained behind in the capsule, and after the removal of the speculum the patient squeezed his lids together, causing escape of vitreous through the incision. In the left eye vitreous humor escaped before the extraction of the lens, so that the latter had to be removed with the spoon. The right eye recovered with but slight inflammatory reaction; the left, however, developed irido-cyclitis. Of the latter (left) organ, it was noted, on October 3d: "Cornea and aqueous cloudy; pupil occluded by exudation masses, and displaced upward; ciliary region painful; abnormal softness of the globe; perception of light." And again, on November 22d: "Tension of eye has become normal; the ciliary region is but slightly sensitive to the touch; a small opening has been cleared through the upper and outer part of the pupil; the patient can count fingers, with this eye, at a distance of three feet." On the day of the patient's discharge, December 1st, no vestige of irritation, sensitiveness, or softness remained in the left eye; the cyclitis had completely vanished, and vision was $\frac{1}{80}$ normal. In the right eye the pupil was clear, and the fundus of the globe distinctly visible, but floating opacities in the vitreous were scattered over the field of vision. These opacities, however, could not be taken as evidence of cyclitis, because the ciliary body in the right eye had not been painful, and the eyeball, after the week first following the operation, had been perfectly free from injection, painless, and of normal tension. It possessed one-eighth of the normal amount of vision. No portion of the iris, in either eye, was included in the cicatrix. The patient was discharged in the forementioned condition.

On January 18, 1876, he returned, with the complaint that, without any external provocation, his *right* eye now suffered. The *left* eye—the one that had been affected with cyclitis—had not been in the least degree painful or reddened during the seven weeks succeeding the discharge of the patient, and on the day of his return showed no trace at all of vascular injection, or of tenderness, on pressure, over the ciliary body; its vision was normal. The right eye, on the other hand, showed all the symptoms of a highly acute irido-cyclitis: intense episcleral injection environed the corneal border, the pupil was plugged with a mass of pus, and displaced toward the place of incision, the globe was soft, the sight was dwindled to a mere perception of light and darkness, and the already spontaneously acute pain became maddening when pressure was made over the ciliary body.

Here an operation, performed for the relief of cataract, had excited *primary cyclitis in the left eye*. The disease, however, had not advanced to atrophy of the globe, but *recovered most perfectly*. About six weeks later, after all the symptoms of the previous cyclitis had disappeared from the *left* eye, the *right* eye, without any external cause, and without any symptoms of the reappearance of disease in the eye originally affected, was visited with an attack of irido-cyclitis, greatly surpassing in severity the primary affection of the first eye. Thus, even *after the complete recovery of one eye from an attack of cyclitis—a recovery not ending in atrophy of the globe—the other eye is not thereby absolutely assured of immunity against an outbreak of sympathetic ophthalmia.*

SECTION III.

PATHOLOGY.

In the preceding section we have considered, so far as is practicable in the preliminary stage of our work, the various individual lesions from which originate the sympathetic diseases usually grouped under the name of *sympathetic ophthalmia*. We now pass to a more accurate description of the manifold forms in which sympathetic ophthalmia appears. The more knowledge we acquire of this class of affections the more multiplied they become. Many forms of ophthalmic disease, whose sympathetic character was formerly and even but recently denied, are now permanently settled in the category of the sympathetic affections; and many others, which are still involved in great doubt, and whose acceptance as sympathetic diseases is properly deferred, may hereafter come to be regarded as integral links in this dangerous chain of maladies.

The following list comprises the sympathetic diseases of the eye: neuralgia of the ciliary nerves; irritation of the retina, and of the optic nerve; functional disturbance of the retina; inflammation, severally, of the conjunctiva, cornea, and choroid; inflammation of the uveal tract, with or without participation on the part of the ciliary body, so that there may be both a sympathetic iritis and a sympathetic choroiditis, without coexisting cyclitis; inflammation of the retina, alone or in conjunction with inflammation of the choroid; inflammation of the optic nerve; glaucoma; disease of the vitreous, and of the lens. Whether all the diseases above enumerated are legitimate occupants of the list of sympathetic affections or not, we shall see in the sequel. We will first describe the symptoms of *sympathetic irritation.*

The *ciliary nerves* play so important rôles in the pathogeny of the sympathetic diseases that, before discussing the subject of ciliary neuralgia, we shall devote a few lines to the anatomical description of these nerves.

The *naso-ciliary nerve* enters the orbit through the sphenoidal fissure, as the third branch of the ophthalmic (sensitive) division of the trigeminus. In the first part of its course it lies on the temporal side of the optic nerve and then passes obliquely over toward the inner wall of the orbit, between the optic nerve and the superior rectus muscle. As it crosses the optic nerve, the naso-ciliaris, having previously given off the long *sensory* root (radix longa) to the ciliary ganglion, sends off from one to three filaments, called the *long ciliary nerves*, which run straight forward to the eyeball. The ciliary ganglion, an oblong flattened body, of about the size of a pin-head, situated between the optic nerve and the external rectus muscle, receives *motor* fibres (radix brevis) from the third cranial nerve (oculo-motor), and *sympathetic* fibres (radix sympathetica) from the cavernous plexus, which surrounds the internal carotid artery. The three roots just mentioned enter the *posterior* border of the ganglion; whilst the *anterior* border gives off the *short ciliary nerves*, which then pass forward to enter the eye. The long and short ciliary nerves split up into fifteen or twenty filaments before piercing the sclerotica around the periphery of the optic nerve, and dividing still further as they advance, run forward, between the choroid and sclerotica, to the ciliary muscle, in which they form a fine net-work, from which numerous fibres are distributed to the iris and cornea. The ciliary nerves, by reason of their triple composition, confer *sensibility* upon the individual parts of the eye, as well as *motility* upon the ciliary muscle, the muscles of the iris, and those of the parietes of the vessels. They are, moreover, probably endowed with *other functions*, which will engage our attention farther on.

In connection with the phenomena of sympathetic irritation, it should be remembered that, when one eye becomes inflamed and painful, from whatever cause, the other can no longer, as a general rule, be used without showing unmistakable symptoms of weariness. In certain inflammations—for example, those phlyctenular lesions of the cornea which accompany the so-called scrofulous affections of the eye—the photophobia of the diseased eye is often propagated to the second, even when the latter is perfectly well, so that *both* eyes are held tightly closed, and are totally incapacitated for use. Or, if the case does not exhibit such extreme symptoms as these, the second eye, in consequence of severe irritation, pain, or inflammation of the first, cannot be employed at fine work without soon becoming tired and strained. Every considerable effort, perhaps for a longer, perhaps for a shorter period, causes the second eye to redden and become irritable, and provokes so painful sensations as seriously to impede its function. Indeed, the presence of a particle of coal-dust in the conjunctival sac of the one eye oftentimes suffices to set up a whole train of symptoms of irritation in the other.

I do not know exactly what name to give to this striking form of "fellow-suffering" (as it were "sympathy") in the well eye. "Sympathetic irritation" is rather objectionable, for, although these words really define the state of things as just described, we feel justified in reserving this expression to indicate a condition which closely borders upon sympathetic ophthalmia, or, indeed, constitutes its preliminary stage. For, while the irritation in the second eye, which is due to pain in the first, usually vanishes with the subsidence of the original pain, or very simple means, such as the application of a compress-bandage to the diseased eye, generally relieves the spasmodic closure of the lids in the other, and enables the patient to separate them freely, this simple form of irritation in the second eye —and here is the main point—*may persist for a long time without danger of involving the organ in substantial lesions.*

On the other hand, where true "sympathetic irritation" is present, we have a very different and infinitely more serious state of matters. For example, an eye that has received an injury, and been very speedily attacked with irritation and inflammation, may excite almost simultaneously, in the opposite eye, so acute and painful phenomena that it is by no means uncommon to hear the patients complain that, for the first day or two after the injury, they were *blind in both eyes*. When the inflammation and pain subsequently subside in the injured eye, the second becomes again quiescent and serviceable, and remains so during a certain interval. After a time, however, without the necessity of any especial exacerbation of the disease in the first eye, and even when the eyeball is no longer spontaneously painful, but only painful or sensitive to the touch, the symptoms of irritation may reappear in the second eye, so that it becomes sensitive when exposed to a brighter light than usual, and fatigued by work that makes but slight demands upon its accommodation. The patient, moreover, may occasionally have noticed, even from the date of the original injury, that the employment of the eye, at the accustomed distance from the work, required a certain effort, which was relieved by holding the work farther from the eye. If the exercise of vision is persistently prolonged, the eye becomes bathed in tears, pain is felt, as well in the neighboring regions as in the eye itself, objects are seen as if through a fog, and if the work be pushed to an extreme limit, the eye becomes utterly disabled for a time. We can, further, often learn by inquiry, that the eye, even when not taxed by exertion, is subject to temporary obscuration of its field of vision. Sometimes, also, during this irritative stage, the patient complains of subjective sensations of light, in the form of sparks or flashes of fire.

It is not probable that these symptoms of "sympathetic irritation" depend, in their early stage, upon textural alteration already present in the eye, for they promptly disappear, once for all, as soon as the opposite eye is enucleated. In those cases in which the symptoms of irritation do not cease in the second eye, notwithstanding the enucleation of the injured eye, but, on the contrary, give place to those of violent inflammation, or in which the inflammation is lit up in the sympathetic eye after the operation, without any preliminary stage of irritation, we must assume that some structural disease, without salient symptoms, had already invaded the second eye at the time when its partner was removed; or that some insidious disease, which did not depend directly upon the disease itself, was on its way toward the second eye, and could not be prevented by the operation; or, finally, that the very operation, practised for the relief of the irritated eye, was itself the cause of the sympathetic ophthalmia.

If *no textural alteration* exists in the second eye at the time of the "sympathetic irritation," the latter must be ascribed to an irritated condition of the ciliary nerves, as well as of the retina and optic nerve. Under such circumstances, it appears to me that the primary involvement is to be sought for *in the retina*, inasmuch as the sensitiveness of the eye to light, the quick exhaustion of the retina by work, the transitory obscuration of the field of vision, and the subjective sensations of light, all point toward this conclusion. This primary irritation or hyperæsthesia of the retina begets a secondary or reflex neurosis in the tract of the ciliary nerves, which consist in great part of sensory filaments from the trigeminus. In consonance with this view, we do not believe that these symptoms depend upon a hidden affection of the muscles, or upon asthenopia of accommodation, such as appears in consequence of the weakness of the muscle concerned in this function. Nor is it our opinion that the holding of the work at a farther distance than usual from the eye is so much a proof that the affection of the ciliary muscle is the primary one, from which the other phenomena of sympathetic irritation proceed, as that the ciliary nerves labor under a reflex neurosis propagated from the primary affection of the retina, so that the contractions of the ciliary muscle, which necessarily provoke pain in the sensory filaments of the sympathetic nerves, are avoided so far as possible.

It is certainly not our intention, in what we have just said, to deny that *primary ciliary neuralgia* may initiate sympathetic disease. This affection, which has its seat in the ciliary and circumorbital branches of the trigeminus, is characterized by violent pain, which is increased by work, so long as work is possible, as well as by light; while, at the same time, the pain does not disappear, even if the patients abandon all exertion on the part of the eyes, and exclude them wholly from the influence of light. Although we cannot discover any definite lesion of the eye, it is evident that the neuralgia of the eyeball is principally located in the ciliary body (the very locality of the chief distribution of the nerves), because even the slightest pressure over the ciliary region exaggerates the pain to an intolerable degree. To diagnosticate cyclitis under these circumstances would be quite unjustifiable, for not a trace of *inflammation* exists in the ciliary body a this period, but simply an exquisitely painful and violent neuralgia of the region involved.

The same irritative condition which has been witnessed in the tract of the ciliary nerves, may also assume a violent type in the retina and optic nerve; so that the symptoms of sympathetic irritation vary according to the functions of the parts involved. The eye affected by sympathy may exhibit the most *intense photophobia*, which, in turn, may develop spasmodic action of the orbicularis muscle, which now presses the eyelids so tightly together that the patient cannot open his eyes at all, and often imagines himself to be blind. Donders has related several cases of this form of severe sympathetic irritation. The fact that the photophobia disappears, and the normal power of vision returns, after enucleation of the opposite eyeball, goes to prove that the spasm of the lids was due to the photophobia alone. We are here to remark, moreover, that the sympathetic irritation of the retina may degenerate not only into intense photophobia, but into the *worst phase of photopsia*, in which the patient is beset with subjective sensations of the most tormenting character. We have already mentioned that the patient may often suffer from transitory sensations of light during the ordinary forms of sympathetic irritation; but it sometimes happens that this phenomenon reaches an extraordinary height, and then constitutes an affection of the most serious importance.

An eyeball is wounded by a penetrating fragment of a percussion-cap. About one year afterward, Alfred Graefe enucleates the injured eye (although its vision is but slightly deteriorated), on account of the distressing subjective sensations in the other eye, which are, however, entirely independent of any demonstrable morbid alteration, while, furthermore, the vision of this eye is absolutely unimpaired. Leber examines the enucleated eye and discovers the fragment of cap adhering firmly to the inner surface of the apparently normal ciliary body. That portion of the retina which covers the ciliary body, and is called the pars ciliaris retinæ, is thickened where it lies applied to the foreign body, and a new formation of connective tissue is found at the *intra-ocular extremity of the optic nerve.* The subjective sensations are not ameliorated by the operation, but reach so extreme a grade that fears are entertained for the life of the patient. A violent degree of photopsia may certainly accompany simple irritation of the optic nerve, but in that case the photopsies vanish after the enucleation of the first eye. Was there not, therefore, in this case of Graefe's, some substantial lesion already present in the sympathizing eye? We shall resume this question in a subsequent place.

But photophobia and photopsies are not the only subjective symptoms of irritation of the optic nerve and retina; for the sympathy may express itself in the form of distinct functional disturbances, or marked *impairment of vision*, without our being able to demonstrate the presence of any definite structural lesion in either the percipient or the conducting apparatus of the eye. We should first mention, in connection with this form of sympathetic irritation, that we may observe not only *momentary* obscuration and limitation of the field of vision, but even longer intervals of suspension of the normal function of the retina. Liebreich gives instances in which the sympathetic irritation of the retina manifested itself by photophobia and obscurations of the field of vision, which lasted from half a minute to a minute, appearing and disappearing at *regular rhythmical intervals*. A still more important form is that sympathetic disturbance of vision which bears some relation to the affection to which v. Graefe gave the name of *anæsthesia of the retina* (proceeding from hyperæsthesia), while Steffan did not hesitate to call it genuine *hyperæsthesia of the retina*. This is the same malady for which Schilling proposed the name of "contraction of the field of vision, without anatomical lesion." This disease is characterized, on the one hand, by a diminution of the acuteness of central vision, and on the other by anæsthesia of the peripheral portion of the retina, so that the field of vision is concentrically contracted, and in a very uniform manner in all directions. The function of accommodation may also be impaired. The ophthalmoscope reveals nothing abnormal, either in the retina or in the optic nerve. Mooren has reported several cases of this form of sympathetic disease, and a case described by Brecht (1874) may here serve for an example.

The injured left eye is very soft at the time of the first examination, but is *entirely free from irritation*. With the right eye, which appears normal, the patient can count fingers, in ordinary light, at a distance of only eight feet. If the eye is fixed upon a given point on a black-board nine inches away, it cannot distinguish the traces of a piece of white chalk at a greater distance than two and a half inches in any direction from the point of fixation. The field of vision is, therefore, concentrically contracted, so that, at a distance of nine inches from the eye, it embraces only a circle two and a half inches in diameter, described around the point of fixation. There are no pathological alterations visible with the ophthalmoscope. After enucleation of the left eyeball, both central and peripheral vision begin to show a decided improvement, and ten weeks after the operation, central acuteness of vision, as well as the peripheral field of vision and the function of accommodation, are all nearly normal. A black splinter of metal is found imprisoned within the enucleated eyeball.

Cohn has reported two cases which, as Leber believes, should be included in the present class of sympathetic affections. We have previously alluded to the pathological changes in eyes that have been subjected to contusions from gunshot wounds (page 49), so that we may here briefly state that the sympathetic disturbance of vision in Cohn's cases was characterized by reduction of central vision, as well as by impairment of the function of accommodation, and, in one of the cases, by severe photopsies, which were repeatedly produced by the most trivial exercise of the eye. Cohn says nothing about the state of the field of vision, so that we do not know whether it was contracted concentrically, if contracted at all. The enucleation of the injured eye completely dissipated the sympathetic troubles. Hyperæsthesia of the retina (not necessarily accompanied by photophobia and photopsia) would appear to be the cause of similar sympathetic disturbances of vision without any structural alterations in the eye.

We now turn our attention from the manifold aspects of sympathetic irritation, to the still more varied forms of *sympathetic inflammation*. In what causal relationship with the inflammation does the irritation stand? Is sympathetic irritation the forerunner of sympathetic inflammation? There is no doubt that the complex of symptoms, characterized by sensitiveness of the eye to light and work, slight transitory congestion of the pericorneal region, painful sensations in and around the eye, and periodical haziness of the field of vision, is to be regarded in the light of a premonitory stage of sympathetic inflammation, which now lies close at hand. It is, however, still an open question, whether the uncomplicated ciliary neurosis, or pure photophobia and photopsia, as well as functional disturbances of the retina without- structural lesions (although these affections can, as a matter of fact, continue, simply as such, for a long time), do not finally become transformed, on the one hand into cyclitis, or on the other into inflammation of the retina or of the optic nerve. It would, however, be incurring a very bold risk to base our therapeutical measures on the assumption that such a state of irritation *never* becomes transmuted into one of inflammation.

In proceeding to consider the different manifestations of sympathetic inflammation, as it affects the individual parts of the eye, we must first notice the cornea. *Sympathetic keratitis* is described by Warlomont as being marked by inflammatory cloudiness of the superficial layers of the cornea, and a profuse development of vessels therein, conjoined with pain in the periorbital region and head, on the affected side, together with intense monocular conjunctivitis. We have already referred to a case in which an eye was destroyed by the thrust of a cow's horn. The eyeball was reduced to a small stump, and, for ten years afterward, remained painless and inoffensive to its mate. After that period keratitis appeared in the second eye, underwent continual relapses during several years, and was rebellious to all treatment until the atrophic stump was finally enucleated, when the sympathetic affection disappeared, as if by magic. In further proof of the sympathetic nature of the disease, it may be stated that an artificial eye, worn after the operation, excited inflammation of the palbebral conjunctiva, with which it came in contact, as well as a fresh outbreak of vascular keratitis in the remaining eye, and that when the artificial eye was thrown aside and poultices were applied to the inflamed cavity for several days, the sympathetic keratitis disappeared without the necessity of having recourse to any other treatment.

Rossander has reported one case of *sympathetic intermittent keratitis;* while Galezowski, Rheindorf, Ledoux and Vignaux have seen cases of *sympathetic kerato-iritis*. Vignaux (1877) observed the latter condition eight times among ninety cases of sympathetic ophthalmia. "In this form of keratitis," writes the last-named observer, "the cornea becomes the seat of a very diffuse (sometimes circumscribed) infiltration, which becomes transformed into superficial ulcerations: while one ulcer heals, another makes its appearance. The iris always becomes implicated in the inflammatory process, and pus is occasionally found in the anterior chamber. The ciliary pain is acute, and the photophobia is almost as excessive as that which we meet with in scrofulous inflammation of the cornea." We must especially notice that the ciliary body does not seem to be affected during these forms of inflammation, which are generally milder than all others. Although not infrequently met with by French writers (constituting as they do almost ten per cent. of Vignaux's series of cases), they are, nevertheless, seldom reported in German medical literature. Perhaps this hiatus has hitherto been due to a lack of vigilance in observation.

Sympathetic ophthalmia may also manifest itself by a genuine attack of *sclerotitis*, unaccompanied by inflammation of the ciliary body. Rossander, for instance, mentions two such cases, in which sympathetic sclerotitis was happily relieved by the enucleation of the injured eye.

Of the various *sympathetic inflammatory processes* that may affect the individual structures of the eye, those which primarily have their seat *in the uveal tract* vastly exceed all others in importance, and they are, further, the ones which most often come under observation and treatment. By reason, therefore, of their great significance and frequency, as well as their destructive effects, it is of the first moment that they should be promptly and accurately diagnosticated, with a view to their timely and appropriate treatment.

Iritis serosa is the least severe of the different forms of sympathetic inflammation of the uveal tract. Suppose that the patient complains of a slight failure of vision in his well eye, whilst the opposite eye, which had, perchance, been destroyed by an injury, is still painful, or, perhaps, only sensitive to pressure. The characteristic symptoms of sympathetic irritation are not present: the worst that the patient complains of is, that for some time past every object has appeared to be covered with a thin cloud. If the medical attendant is not alert, the actual pathological process may be overlooked, and perhaps mistaken for a sympathetic functional disturbance of the retina. Careful investigation, however, by daylight, or by the oblique illumination of the eye (the image of a lamp-flame being projected upon the cornea by a strong convex lens), will reveal small, grayish, punctated opacities on the posterior surface of the *lower half* of the cornea, while, if the pupil be illuminated by the ophthalmoscope (the patient looking downward), its area will appear to be filled, as it were, with fine dust, interspersed here and there with small, dark specks, varying in size from a pin-head to almost microscopical minuteness. It may, indeed, happen that with the unaided eye, or even with oblique illumination, nothing unusual can at first be discovered, and that it will require the use of the ophthalmoscope before the punctated appearance of the cornea can be accurately recognized by the incident light. We shall, moreover, now begin to notice that although the eye had been pale before the examination, the irritation incident to this procedure has of itself sufficed to provoke a faint rosy zone of episcleral injection around the margin of the cornea. We shall also, perhaps, see that the pupil, although perfectly free, and nowhere adherent to the anterior capsule, does not react so promptly to the influence of light and shade, as when in a normal condition, and that, although a comparison with the other eye may not now be practicable, the pupil is evidently rather larger, and the anterior chamber much deeper than in the mate. Sensitiveness of the ciliary body is not necessarily educed by pressure. The tension of the globe is, on the whole, quite normal: sometimes it may be increased, but it is never diminished. Such, then, are the most simple indications of serous iritis.

We have already mentioned that the fine opacities in iritis serosa are situated on the posterior surface of the cornea. We assume that there is an increased exudation of serum (with the addition of pus-corpuscles and coagulable material) from the iris into the anterior chamber, which latter is consequently deepened, owing to the pressing backward of the iris and lens by the superabundant fluid. The pus-corpuscles and small masses of coagulable lymph gravitate downward, and become deposited on the posterior surface of the cornea; so that we need not be at all surprised at the general absence of these "precipitates" on the upper

4

portion of the cornea. The presence of these punctiform deposits is pathognomonic of iritis serosa. Although, strictly speaking, they are not always deposits precipitated from the aqueous humor, nevertheless, the difference in their origin does not alter their diagnostic value. If we puncture the anterior chamber and catch in a watch-glass a portion of the contents, together with some of the precipitates upon the posterior surface of the cornea, we may experimentally convince ourselves that these opacities are, as a rule, actual deposits, consisting of particles of coagulated fibrin, enclosing pus-corpuscles in greater or less number. On the other hand, it has been found, during histological investigations, that these punctiform spots on the posterior surface of the cornea may also be caused by inflammatory changes in the epithelial lining of the membrane of Descemet, and even in the posterior laminæ of the proper corneal substance. It need not, therefore, surprise us that these "precipitates" should now and then be observed, not only on the lower portion of the posterior corneal surface, but also opposite the pupil, and sometimes even scattered over the upper half of the cornea. Nevertheless, true inflammation of the membrane of Descemet, or genuine keratitis postica, is always to be regarded as characteristic of the serous form of iritis, inasmuch as it is directly excited by the morbid and irritating contents of the anterior chamber. It is chiefly the accompanying turbidity of the aqueous which causes the hazy appearance of all objects seen with the affected eye.

It is important to note, in this connection, that while sympathetic *iritis serosa* usually appears under the unobtrusive symptoms above described, those forms of this affection which are *independent of a sympathetic origin*, are wont to be more distinctly and prominently expressed. In the latter, we not unfrequently notice very marked pericorneal injection, extreme deepening of the anterior chamber, and, instead of the *fine* punctated exudation on the membrane of Descemet, coarse, grayish, or even yellow, nodules, as large as pin-heads or hemp-seeds. It should, moreover, be distinctly borne in mind that we are not directly to diagnosticate iritis serosa, on account of the presence of nodules of exudation, but to look about for other alterations in the eye. If, for example, we have simultaneously, an inflammatory adhesion of the margin of the pupil to the anterior capsule of the lens, it would be wrong to call the case one of iritis serosa. The precise difference between a serous and a plastic iritis lies in this fact, that in the serous form there is not a sufficient degree of plastic inflammation to effect any such adhesion between the edge of the pupil and the capsule. On the other hand, however, it is by no means uncommon, in a case of violent iritis plastica, to observe flocculent masses of pus or lymph floating about in the aqueous humor, as well as considerable proliferation of the epithelial cells of the membrane of Descemet.

It is further important for us to insist upon an accurate discrimination between the *plastic* and the *serous* form of sympathetic iritis. *Sympathetic iritis plastica* closely simulates, at the outset, *common* plastic iritis, which, as a rule, leads to only partial adhesions of the pupillary edge to the anterior capsule, but not to a marked agglutination of the posterior surface of the iris to the capsule of the lens. Sympathetic iritis plastica is, on the contrary, very prone to develop into that more severe grade of iritis in which the adhesion rapidly involves the whole circumference of the pupillary border, so as to shut off all communication between the anterior and posterior chambers, producing the condition technically termed *exclusion* of the pupil. Under these circumstances, the central portion of the anterior capsule, opposite the pupil, may still remain perfectly clear, or, at the most, be covered with so scanty a morbid product as not essentially to obstruct the passage of the rays of light. When, on the other hand, the pupil is filled with a thick pseudo-membrane, or even with a dense plug of exudation, so that the pupillary area is completely abolished, the condition is called *occlusion* of the pupil. As the exclusion of the pupil may exist without its occlusion, so, conversely, occlusion may not necessarily involve exclusion. For it is easy to comprehend that a false membrane may wholly cover the pupil without necessitating a continuous adhesion between the entire circumference of the pupil and the anterior capsule; so that, at one point or another, beneath the edge of the membrane, an opening, however small, may still remain, and so preserve the communication between the two chambers.

Occlusion of the pupil, although obstructing the passage of the rays of light, may cause no real damage to the eye itself; but *exclusion* of the pupil, while presenting no direct barrier to the vision, very frequently destroys the affected eye. We may conceive that the aqueous humor is secreted by the ciliary processes and iris, or perhaps only by the posterior surface of the latter. We know, besides, that the aqueous normally finds its way out of the anterior chamber, by filtration and diffusion into the veins immediately adjacent to its periphery. If, now, the communication between the anterior and posterior chambers is abolished by exclusion of the pupil, the fluid secreted into the posterior chamber, from the ciliary processes and the posterior surface of the iris, is deprived of its normal means of escape into the anterior chamber, and then into the pericorneal veins, as well as into the sinuses of the ligamentum pectinatum iridis, so that an abnormal accumulation of aqueous takes place in the posterior chamber. It happens, therefore, as soon as the pressure of the fluid in the posterior chamber exceeds that in the anterior, that the inequality manifests itself by the bulging forward of the iris into the anterior chamber, except at those points where it is held back by the adhesions. The protrusion forward of the periphery of the iris, accompanied by a crater-like depression of its pupillary edge, is, therefore, a sign of exclusion of the pupil. So long as this phenomenon is absent we cannot diagnosticate exclusion of the pupil; for, even with the assistance of mydriatics, we are unable to declare positively that some minute hole does not exist, at one point or other, around the apparently completely adherent margin of the pupil.

Now, this imprisonment of the aqueous humor behind the iris, with the jutting forward of the periphery of the latter membrane, almost invariably leads to a complex set of symptoms, which are comprised under the name of *secondary glaucoma*, in which, with more or less violent attacks of inflammation, the tension of the eye increases and vision diminishes; or the globe remains hard, while vision gradually decreases to utter blindness, *without any intercurrent inflammatory phenomena whatever*. The extinction of vision then depends upon a lesion of the optic nerve, producing its total atrophy. Glaucoma is that affection of the eye which, with evident hardness of the globe, and with or without inflammatory exacerbations, leads to blindness. When the glaucoma depends upon some affection previously present in the interior of the affected eye—as in our case, for example, upon a bulging iris, produced by accumulation of fluid behind it—the disease is called secondary glaucoma. It follows, therefore, that secondary glaucoma may sometimes occur 'in a sympathetically diseased eye, and cannot *always* be regarded as a part of the sympathetic process. For when sympathetic iritis plastica is followed by continuous circular adhesions (exclusion of the pupil), and finally produces secondary glaucoma, the latter disease depends wholly upon the adhesions, and not at all upon the sympathetic origin of the latter.

We should here incidentally remark that an inclination prevails, whenever sympathetic ophthalmia is met with, to diagnosticate a *cyclitis;* or when the signs of a *plastic* cyclitis are wanting, to find, at least, a *serous* cyclitis. But we are not of those who believe that the bulging forward of the periphery of the iris, in sympathetic ophthalmia, or in secondary glaucoma, furnishes sufficient ground for inferring the existence of any sort of cyclitis, inasmuch as an analogous condition of the iris may likewise be developed in common inflammations of this membrane (which are quite independent of any sympathetic foundation), without properly exciting any suspicion of even serous cyclitis. The idea of assuming the presence of cyclitis, in the generality of cases of sympathetic ophthalmia, is just as unnecessary as the possibility of establishing the fact of its presence is questionable.

The mildest form of sympathetic disease of the uveal tract is *serous iritis; plastic* iritis comes next in order of severity, chiefly on account of the annular, posterior synechiæ, or exclusion of the pupil, to which it is so prone to give rise; but incomparably the most serious manifestation of sympathetic uveal disease is the so-called *iritis maligna*, which is nothing else than a plastic irido-cyclitis. In iritis serosa, adhesions do *not* commonly take place between the iris and anterior capsule; in plastic iritis adhesions occur, but they are as a rule, limited to the *pupillary border* of the iris; whilst iritis maligna is characterized by *extensive* agglutination of the posterior surface of the iris to the anterior capsule of the lens. Inasmuch as, in iritis maligna, choroiditis is almost always superadded to the irido-cyclitis, and the integrity of the retina becomes thereby threatened, sympathetic uveïtis attains, in iritis maligna, its culminating degree of severity. For when the iris, ciliary body, and choroid are all involved in the inflammatory process, the eyeball is usually consigned to atrophy.

It is not necessary for us at this point to sketch the symptoms of *sympathetic iritis maligna*, inasmuch as we have already (pages 26, 27, and 28) clearly described irido-cyclitis, as well as irido-cyclo-choroiditis, of the primarily affected eye, as they occur either spontaneously or in connection with injuries; and the sympathetic forms do not differ materially from the primary, except in the more frequent opportunities we have for observing the former. In other words, the genuine form of the disease in question is much oftener seen in the eye affected by sympathy, than in the eye originally affected, in which latter the regular type of the disease is frequently obliterated by the immediate effects of the injury.

What *relationship* and mutual dependences do we find *among the different forms of sympathetic iritis?* What are their course and issue? It is true that iritis maligna is more frequently met with than the serous or the plastic form of iritis; nevertheless, the two last-named species of this malady are not so rare as is commonly supposed. Statistical inflammation touching the comparative frequency of iritis serosa is not easily obtainable, because the great majority of individuals who are affected with this variety of sympathetic disease certainly do not come under the notice of a medical attendant. It may be inquired how this is possible? Is not serous iritis merely a forerunner of the more important kinds of inflammation of the iris? Is it not the pioneer of iritis maligna? We must promptly answer this question in the negative. Then, again, if the serous form of iritis *were* transmutable into iritis maligna, we should probably find few opportunities to observe the former, for the reason that only the severer grades of iritis are likely to bring the sufferer under professional observation. The recognition, therefore, of sympathetic iritis serosa, as a distinct affection, is not, in some respects, of great practical moment. It is, however, of importance for us to know that iritis serosa has no inherent tendency to lapse into the worst forms of iritic inflammation. Whenever a surgeon enucleates an injured eye, on account of sympathetic serous iritis, and, upon subsequently seeing amelioration of the symptoms, flatters himself that his well-timed interference has happily prevented a sympathetic plastic irido-cyclitis, and blindness of both eyes, he has, in all probability, been the victim of a self-pleasing error. However, we do not here desire to anticipate a discussion of the indications for enucleation, but only parenthetically to remark, that iritis serosa has nothing in common with iritis maligna, and, as a very general rule, runs a favorable course without extirpation of the eye first affected; and furthermore, that when a case of sympathetic iritis serosa has degenerated into a worse form of iritis, after the enucleation of the first eye, *the operation itself has, in all probability, been the cause of the new sympathetic process.*

The relationship which exists between iritis plastica and iritis maligna calls for some comment. It is very generally stated in connection with iritis maligna that adhesions between the greater portion of the posterior surface of the iris and the anterior capsule of the lens need not be present in order to establish the diagnosis, but that, in the beginning, the adhesion may be limited to the pupillary border, while the periphery of the iris is, at the same time, bulged forward by the serum confined behind it. It is further averred that at a later stage of the affection the serous gives place to a plastic exudation, which then firmly and extensively glues together the iris and anterior capsule, and, by subsequent contraction, retracts the periphery of the iris. I will here place no significance upon the fact that I have never, in my personal experience, witnessed this transition from a protrusion to a retraction of the periphery of the iris; but I must openly say, that when I see total circular posterior adhesions, with bulging of the periphery of the iris, in a case of sympathetic ophthalmia, I do not think of diagnosticating iritis maligna, but only the common form of plastic iritis with exclusion of the pupil, especially as the tension of the eyeball so affected is not diminished, but is either normal or augmented. Such an iritis, *if secondary glaucoma should not supervene*, might run a relatively favorable course. Nevertheless, I do not like to take the risk in such cases, but let the bulging of the iris be to me the signal for surgical interference. It is quite a matter of course that errors of diagnosis may sometimes occur in these cases, for the iris may not only be thrust forward by the aqueous humor imprisoned in the posterior chamber, but likewise, by extensive *plastic* exudation in the same locality, as I was once convinced upon dissection of an eye. In the case here instanced, it was easy to see how the iris might have first been bulged forward, and then retracted at its periphery by the shrinking of the exudation.

The course of iritis maligna varies according to the different structures involved in the inflammatory process. Sometimes it is almost wholly confined to the iris and ciliary body, so that the integrity of the vitreous and choroid (and consequently of the retina), is mostly spared. The eye, under the latter condition of things, retains perfectly or tolerably well its normal tension (even when the inflammation has covered the pupil with a pseudo-membrane), is frequently promptly sensitive to light and shade, and in cases where the pupil remains clear, or is obstructed by only a thin film, preserves a corresponding degree of vision. The majority of cases of iritis maligna, however, terminate in atrophy of the globe, on account of the consecutive inflammation of the choroid, so that perception of light is either totally extinguished, or reduced to an insignificant amount.

In the form of sympathetic ophthalmia now under consideration (plastic irido-cyclitis), we sometimes notice a remarkable phenomenon, which is of great value in connection with the *pathogeny* of this, as well as of other sympathetic affections in which it occurs, and which will, therefore, be further discussed in another place.

It consists of the manifestation of pain, either spontaneously or on pressure, at a spot on the sympathetic eye, corresponding *symmetrically* to a point on the injured eye, which is still spontaneously painful, or painful only to the touch. If, for illustration, the most painful place of the eye first affected is situated near the upper and outer edge of the cornea, perhaps at the spot where a scleral wound, with incarceration of a portion of the iris, has occurred, the chief or even exclusive seat of pain in the second eye will likewise be located at a precisely corresponding point on the supero-temporal margin of the cornea.

In the present relation another phenomenon which has been observed in several cases of sympathetic cyclitis deserves mention. Schenkl discovered several silvery-white eyelashes on the temporal half of the upper left eyelid of a boy, nine years of age, at a time when this eye was sympathetically inflamed, in consequence of an injury received by the right eye. On the upper lid of the *right* eye *all* the eyelashes were perfectly white, with the exception of a minute portion of their extremities, which was very dark. Jacobi also noticed in an eye, sympathetically affected with irido-cyclitis, that the lashes of the nasal half of the upper lid were altered in color to snow-white, whilst on the outer half of the same lid the lashes were black and white in about equal proportions, the lower lid presenting merely a few white hairs.

Have we now exhausted all the forms of sympathetic disease that may invade the *uveal tract?* It would seem not. Let us first notice a case reported by Horner (1873). In an eyeball which has long concealed a foreign body, symptoms of irido-cyclitis set in. In the opposite, heretofore healthy, but somewhat myopic eye, a rapidly progressing impairment of vision takes place. The ophthalmoscope reveals, in explanation of the latter defect, a *peculiar form of patches in the choroid*, chiefly in the neighborhood of the macula lutea. Very numerous, minute, yellowish white, imperfectly defined specks, are seen behind the retina. The disease progresses painlessly and without signs of irritation. The spots of exudation, in the choroid, enlarge and coalesce. After a year vision has become so much reduced that fingers cannot be counted at a greater distance than four feet with the central portion of the retina, and seven feet with excentric vision. The function of the retina suffers, in this case, in consequence of the extension of the choroidal exudation to the layer of cones at the yellow spot. There was no well-defined primary sympathetic affection of the retina.

Vignaux (1877) discovered, with the ophthalmoscope, a commencing *atrophic choroiditis* of sympathetic origin, which was the cause of a very pronounced disturbance of vision.

The conjunction of choroiditis with retinitis (*cho-*

roido-retinitis) as a form of sympathetic ophthalmia, was described by v. Graefe in 1866; although, according to the statement of Laqueur, a sympathetic neuro-retinitis had been previously noticed by Rheindorf (1864). After the extraction by v. Graefe of a dislocated chalky lens from the anterior chamber, cyclitis ensues in the same eye. Six weeks after the operation, the sight of the other eye, which has hitherto been perfectly good, begins suddenly to be impaired, although no pain is noticed. The ophthalmoscope discloses a delicate and diffuse cloudiness of the retina all around the entrance of the optic nerve. Soon afterward, slight symptoms of iritis serosa are noticed, in the form of very delicate punctiform opacities in the membrane of Descemet. After vision has sunk to one-eighth of the normal amount, and the disease has continued at its acme for several weeks, a gradual but uninterrupted improvement takes place. The morbid appearances visible with the ophthalmoscope recede less rapidly than the functional disturbances. Disseminated patches of exudation are conspicuous on the choroid, for a considerable time, while the fine punctiform deposits on the posterior surface of the cornea are the slowest to disappear. The field of vision is complete in every direction, and vision is increased to four-fifths normal.

In the second of v. Graefe's cases, a patient, twenty years of age, blind in one eye since childhood, complains that the ruined eye has been painful during the last few months. The globe of the best eye is moderately sensitive to the touch, and there is some impairment of vision. A slight haziness is diffusedly spread through the retina, circumscribed opacities are seen in the vitreous, and the choroid exhibits trivial alterations of structure. After enucleation of the blind eye, the sympathetic manifestations slowly disappear from the other.

Schweigger (1875), however, alludes to the foregoing diagnoses of v. Graefe only to throw doubt upon them, and adds that it requires a number of analogous cases to supply satisfactory evidence of the correctness of such a diagnosis (sympathetic retinitis). For that reason we must here notice similar cases.

Pooley (1871) reports two cases of sympathetic ophthalmia, distinguished by the occurrence of neuro-retinitis. In both of them the injured eye was still abnormally sensitive; whilst iritis, and molecular opacities in the vitreous, were conjoined with the retinal affection, in each case. Galezowski (1871) diagnosticates sympathetic retinitis, characterized by whitish exudations and hæmorrhagic extravasations into the retina, followed by recovery, but with permanent obliteration of some of the implicated vessels. He supports his diagnosis by a similar case of Dolbeau's, which he observed with the latter. Gosselin (1872) speaks of a case of sympathetic inflammation of the retina and choroid, marked by pigment spots, ecchymoses, and inflammatory exudations, together with a small posterior adhesion. The vision of the sympathizing eye became suddenly impaired, at a time when the stump, to which the opposite injured eyeball had become reduced, was the seat of an unusual exacerbation of pain. H. Müller (1873) relates that Jacobson saw a sympathetic choroido-retinitis localized in the neighborhood of the entrance of the optic nerve, the other eyeball being at the time in a state of painful atrophy, ensuing upon cyclitis produced by a cataract operation. Hirschberg (1874) recognizes a sympathetic retinitis, characterized by great hyperæmia of the retinal veins, together with slight diffuse cloudiness of the retinal structures, at a period when the opposite phthisical eye was still very painful to the touch over the ciliary region. Pflüger (1875) meets with sympathetic symptoms in the form of inflammation of the intra-ocular extremity of the optic nerve and the circumjacent portion of the retina. We have already mentioned this case, in which it was found, upon dissection, that an inflammatory infiltration of the iris, unaccompanied with cyclitis, was the lesion affecting the primarily diseased eye. Among the ninety cases adduced by Rossander (1876), sympathetic choroido-retinitis figures three times, although one of these cases holds its position with doubtful propriety, according to the opinion of Rossander himself. In

Leber's work (1877), "Ueber die Krankheiten der Netzhaut und des Sehnerven" (On the Diseases of the Retina and Optic Nerve), only a single paragraph is devoted to sympathetic retinitis. "The affection;" says Leber, "is usually conjoined with serous irido-cyclitis and haziness of the vitreous; after the media clear up the ophthalmoscopic evidences of the affection are sometimes unmistakable." The sympathetic retinitis is usually characterized by a diffuse cloudiness of the structures of the retina, to which a redness of the disc of the optic nerve is usually superadded. But, according to Leber, the retinitis is not simply associated with irido-cyclitis, but is dependent upon the latter; for he commences by saying that "sympathetic irido-cyclitis also leads, now and then, to the development of retinitis." Finally, Vignaux (1877) narrates several cases of sympathetic choroido-retinitis, as well as of retinitis, without iritis or irido-cyclitis. In some of the latter cases the ophthalmoscopic changes are described so meagrely as to throw doubt upon the positive presence of either choroiditis or retinitis, and the sympathetic affection in these cases might as well, or better, be accepted as amblyopia without underlying structural changes. Nevertheless, the existence of retinitis, as an expression of sympathetic disease of the eye, can no longer be regarded as an open question. This kind of retinitis is very generally characterized by diffuse cloudiness of the retina; but whether the sympathetic nature of such forms of retinitis as Galezowski and Gosselin describe, is to be established rather by the presence of other and deeper changes in the retina, cannot to-day be decisively settled.

We should here notice a certainly very important point in connection with sympathetic retinitis. Schnabel (1876) has stated (and Leber has likewise expressed a similar opinion) that common iritis is frequently complicated with diffuse retinitis. If, therefore, retinitis does not really appear as an independent sympathetic affection, but is only superinduced upon sympathetic iritis, the sympathetic character of the affection fails as absolutely as does that of secondary glaucoma, when the latter malady supervenes upon a complete posterior synechia of the pupillary margin of the iris, resulting from sympathetic iritis. Notwithstanding the occurrence of this complication, however, there is no doubt that retinitis, *without* iritis and cyclitis, may arise in a wholly independent manner, from sympathy with the offending eye. I go even farther, and say: the frequent presence of irido-cyclitis, interfering with the employment of the ophthalmoscope, prevents the clinical establishment of the fact that retinitis is a very common manifestation of sympathetic disease; or, in other words, that *many more cases of retinitis are sympathetic than those in which clear and unmistakable evidence of the fact can be obtained.* The last suggestion is of importance in connection with the pathogeny of the sympathetic diseases, and we shall have occasion to resume it farther on.

We leave the sympathetic diseases of the retina with the remark that the case of typical pigment-degeneration of the retina (*retinitis pigmentosa*), described by Robertson (1871) as a sympathetic affection, was manifestly connected (Leber) with a binocular disease, which existed previously to the injury to which the supposed sympathetic disease was attributed.

We now pass into an uncommonly dark province, viz., that of the *sympathetic affections of the optic nerve.* Sympathetic retinitis may, as we will here at once state, be propagated to the second eye along the path of the optic nerve; but is the same statement applicable to the other diseases of the optic-nerve tract? Dransart has added much to the description of this subject: but we shall only mention his assertion that simple atrophy of the optic nerve is to be ranked as one of the sympathetic affections. But he certainly weakens his statement very much when he includes "atrophy of the choroid, posterior synechiæ, and cataracts" among the "frequent accompaniments" of sympathetic atrophy of the optic nerve. Mooren saw a case in which atrophy of the optic nerve of one eye, caused by a contusion, was followed by atrophy of the optic nerve of the opposite eye. This last case is clearly entitled to be called an example of sympathetic disease, in so far as every affection is to be regarded as sympathetic, the reproduction of which in the second eye is ascribable only to a pre-existent disease in the first eye. The question, however, of practical significance is: Whether we can have simple sympathetic atrophy of the optic nerve in the second eye, under the same circumstances in which other sympathetic affections generally become developed? I would not like to deny off-hand the possibility of the occurrence of such a phenomenon. Indeed, from my personal observation of two somewhat enigmatical cases, I cannot wholly avoid the belief that we may occasionally discover the ophthalmoscopical picture of *simple atrophy of the optic nerve*, which is directly of sympathetic origin.

We have already alluded to the danger of implication of the second eye which now and then attends the enucleation of the first eye, when performed for prophylactic purposes, and it is now our purpose to describe the sympathetic phenomena which are sometimes seen in the second eye after the surgical removal of its mate. Colsmann (1877) removed an eyeball which had atrophied in consequence of an injury, and was ominously painful.. A few days after the operation, the acuteness of vision in the remaining eye sank to one-seventh of the normal amount. Three days later the ophthalmoscope revealed distinct cloudiness of the optic disc and of adjacent parts of the retina, the cloudiness being especially conspicuous in the vicinity of the yellow spot. The field of vision was at the same time concentrically contracted. Under appropriate treatment pursued for six months, vision became normal and the visual field complete in every direction. Colsmann also reported a second case of the sort, from Mooren's clinic. A few months after the prophylactic removal of an injured eyeball, the patient complained of subjective flashes of light in the remaining eye, but vision was still normal. Six months later, the acuteness of vision was exceedingly diminished, the patient only being able to read print the size of No. 19 of Jaeger's test-types (one and one-half to two centimetres in height). Inflammation of the optic disc, with very extensive cloudiness of the retina, was discovered with the ophthalmoscope. The final result of this case is not known. Colsmann states that Hugo Müller had, at an earlier date (1873), described a case in which, five days after the removal of a degenerated and enlarged eyeball, the patient, without previous symptoms of sympathetic disease, began to complain of the periodical envelopment of the whole field of vision with a shining white cloud, accompanied by subjective sensations of light. In the intervals of these attacks, no impairment of vision could be ascertained, but the retina was cloudy in the neighborhood of the optic papilla. Later, however, without change in the ophthalmoscopic appearances, the power of vision began to deteriorate rapidly, but was restored after a course of treatment consisting of local abstractions of blood and the administration of mercury. We must not forget to add that, several months afterward, the patient experienced an attack of cyclitis, with increase of intra-ocular pressure (sympathetic glaucoma?), which was successfully treated by iridectomy.

We are here led to seek an answer to an important question: *Is there a sympathetic glaucoma?* The question is not whether a sympathetically diseased eye can lose its sight while laboring under the characteristic symptoms of glaucoma (the glaucomatous symptoms being, in such a case, simply superadded to those of the sympathetic disease), but it is whether primary glaucoma can be developed in the *second* eye, solely from sympathy with the eye first diseased. In other words, can a disease, whose symptoms, briefly expressed, are persistently increased tension of the eye, pulsation of the central vessels of the retina, and an affection of the optic nerve usually characterized by excavation of its intra-ocular extremity, arise directly from a disease or injury of the other eye, and continue, with or without inflammatory phenomena which have their seat in different parts of the eyeball, until the sight of the affected organ is destroyed?

Still another limitation must be made. It sometimes happens, after the operation of iridectomy has been performed for the relief of glaucoma of the one eye, that the other, hitherto perfectly healthy eye, is attacked with the most violent symptoms of acute glaucoma, so that the patient, upon whom the operation on the first eye was, perhaps, undertaken merely for the removal of pain, and with no hope of restoring its lost sight, becomes totally blind. The question whether, under these conditions, the outbreak of glaucoma in the second eye is of sympathetic origin, and ensues upon the operative injury inflicted on the first eye, in the same mode in which sympathetic disease may proceed from any other kind of traumatic injury of the organ, is here answered in the negative, its fuller discussion being postponed until we publish our work on the theory of glaucoma.

Let us reduce our statement and inquiry to the following terms: An eye is destroyed by irido-cyclitis, and the opposite eye becomes, in consequence of the first lesion, affected with sympathetic serous iritis. Every serous iritis, of whatever origin, may *possibly* cause secondary glaucoma. I have never personally seen this effect produced by sympathetic serous iritis; but, even admitting its occurrence, the fact is beside our question. Then, again, instead of serous iritis, the sympathy may manifest itself in the shape of plastic iritis, which may excite secondary glaucoma by the roundabout way of exclusion of the pupil. We cannot deny that this complication really may occur in the sympathetic eye, but the admission does not answer our question, which is: Can *primary* glaucoma be sympathetically produced in the second eye by an irido-cyclitis, or an irido-cyclo-choroiditis of the first eye?

Sympathetic glaucoma appears to have been first described by v. Graefe (1857). After narrating a particular case, he superadds the remark that he has "repeatedly met with a similar condition of things, viz.: absolute amaurosis of one eye, due to the destructive effects of choroiditis; and amblyopia of the other eye, without any signs of irritation whatever, although the affection was accompanied with progressive limitation of the field of vision, as well as *excavation of the optic nerve*, visible with the ophthalmoscope." V. Graefe thought it possible that "disturbance in the circulation and secretion of the choroid might cause increased intra-ocular pressure and consequent cupping of the optic nerve entrance;" in other words, a true sympathetic glaucoma. Many other published accounts of sympathetic glaucoma are extant (Horner, Mooren, Coccius, Carter, H. Müller, Pomeroy, Rossander, Vignaux); and divers authors who have, perhaps, no personal knowledge of sympathetic glaucoma, accept it on the ground of v. Graefe's early observations. Nevertheless, this form of sympathetic ophthalmia falls somewhat short of general recognition. Maats (1865) refuses to concede it, and Brecht (1874) expresses his opinion that in v. Graefe's cases the supposed affection was mistaken for sympathetic amblyopia with limitation of the field of vision, without alterations of structure. But the most powerful antagonist of v. Graefe's observations is v. Graefe himself. For in 1866, in connection with his first description of sympathetic choroido-retinitis, he emphasizes only two forms of sympathetic inflammation, viz., iritis maligna and iritis serosa, and positively asserts that sympathetic irido-cyclitis "*never*, or only in the rarest exceptional cases, shows any tendency to produce an increase of the intraocular pressure, or an excavation of the optic nerve."

It now seems doubtful whether typical simple *glaucoma without inflammatory symptoms*, can be unconditionally admitted into the group of sympathetic affections, especially since v. Graefe himself abandoned this theory, which he at first constructed upon the basis of a few cases which seemed to support it. I would further suggest that there is a manifest inconsistency in acknowledging the existence of this kind of sympathetic glaucoma, so long as it continues to be regarded as a *secondary glaucoma* following serous cyclitis. For the presence of serous cyclitis would, under the latter restriction, only be revealed by the glaucomatous symptoms; and in case the glaucoma were viewed simply as a product of serous cyclitis, the very nature of a primary sympathetic glaucoma would be prejudiced. Primary glaucoma simplex would then be nothing else than a serous cyclitis; but to designate as a *primary sympathetic* glaucoma, a *secondary* glaucoma resulting from serous *cyclitis*, would be quite inadmissible.

The existence, as a sympathetic affection, of *acute glaucoma*, *i.e.*, primary glaucoma *with all its peculiar inflammatory phenomena* (which we shall not stop to describe in this place), must be regarded as extremely problematical, and as not hitherto satisfactorily demonstrated. Even the case reported by Jany (1877), who saw the right eye affected by what he supposed to be sympathetic acute glaucoma, during an attack of scleritis and iritis of the left eye, is lacking in some of the indispensable characteristics of a sympathetic disease. But the case is quite different, where increase of tension is superadded to those inflammatory symptoms which are diagnostic of iridocyclitis. Even where increase of intraocular pressure is noticed in connection with ciliary injection, sensitiveness of the ciliary body to the touch, adhesions between the iris and anterior capsule, and opacities of the vitreous, glaucoma is not necessarily present, and certainly not a sympathetic glaucoma. *Augmented intraocular pressure may be present during every acute inflammation of the eye, of whatever kind or origin.* But if the increased intraocular pressure, under the influence of which vision is sooner or later destroyed, is not permanent, although it may be variable, the disease is *not* glaucoma. The heightened intraocular pressure, which may be present at one stage in irido-cyclitis, subsides in the generality of cases; but even if this were not the case—if the eyeball remained abnormally hard until vision were destroyed—the case would evidently be one of secondary glaucoma, ensuing on irido-cyclitis. The inflammatory symptoms of irido-cyclitis differ so widely from those of glaucoma, that there can be no risk of mistaking a primary glaucoma for an irido-cyclitis. It is the irido-cyclitis, and not the secondary glaucoma, developed from it, which is the sympathetic affection.

A very peculiar form of sympathetic glaucoma, called *sympathetic hæmorrhagic glaucoma*, was described by H. Pagenstecher. (1877). Hæmorrhagic glaucoma is characterized by the extravasation of blood into the retina, accompanied by the most violent symptoms of glaucoma, so that the disease has sometimes been called a secondary glaucoma. According to the description given of Pagenstecher's case, however, the glaucomatous phenomena were *first* noticed, and subsequently followed by the retinal effusions. The left eye, from which the sympathetic affection in the opposite eye was supposed to proceed, showed at the time when its partner was affected nothing more than an ulceration of the cornea, which had not yet caused perforation. Later, a perforation of the cornea ensued, and led to phthisis of the globe. At the date of the enucleation of the left phthisical and blind eye, its tension was augmented; it was only moderately sensitive to *heavy* pressure (consequently less sensitive than a healthy eyeball), and the cornea, which was flattened, and mostly converted into cicatricial tissue, was extremely anæsthetic. The same anæsthetic condition was noticed in the conjunctiva. The operation was followed by a decided improvement in the condition of the right eye, which, however, again became worse several weeks after the enucleation, during the course of a lobular pneumonia. It again improved; but, in consequence of the passing of the patient from observation, the case was not followed to its conclusion. Can any positive causal connection between the diseases of the two eyes be here made out? The improvement of the abnormal tension and impaired vision, which followed the enucleation is very striking, and favors this view. But did not the rest and restricted diet (to which the "plethoric sexagenarian, who was not averse to the pleasures of the table," must certainly have been submitted, for a time at least, after the operation) have an influence in producing the (possibly only transitory) change for the better? Certainly, the condition of the primarily diseased eye, as well at the time of the first "sympathetic" glaucomatous attack of the right eye, as at the time of the enucleation, was not such as to establish beyond a doubt its agency in exciting the disease of the second eye.

To fill the complete catalogue of sympathetic diseases, we will further mention that Schmidt (1874) discovered a few *opacities pervading the vitreous*, in the form of grayish-black filaments, which he ascribed to a sympathetic source. There was no trace of accompanying iritis, nor of other inflammatory processes in the uveal tract.

Finally, Brière (1875) reports a case of *sympathetic cataract*. The opinion expressed by Brière, however, that the cataract described by him should be grouped among the sympathetic affections, is arbitrary. A well-authenticated case of sympathetic cataract remains for future discovery.*

The severest forms of sympathetic disease are inflammations of the iris, the ciliary body, and the choroid, on the one hand, and those of the optic nerve and the retina on the other. The serious lesions of the latter structures are usually concealed by the inflammatory processes that simultaneously occur in the uveal tract. Among the sympathetic affections of the uveal tract, iritis serosa constitutes a remarkable exception to their generally dangerous character. It sounds paradoxical, but it is nevertheless true, that the existence of *sympathetic serous iritis* need excite less anxiety than that of *sympathetic irritation*, for the

---

* Krückow (1880) has, however, described two cases, in which the sympathetic cataract revealed itself, in each instance, in the form of an opacity, confined exclusively to the anterior capsule of the lens.—TRS.

latter affection frequently sets on foot the worst forms of sympathetic ophthalmia, proceeding to the destruction of the eye; while genuine simple iritis serosa possesses very little inherent tendency to destructive results.

Sympathetic ophthalmia is especially prone to be caused by injuries of the eye, because those morbid processes which constitute it are much more frequently of traumatic than of spontaneous origin. Modern ophthalmology, instead of diminishing the sources of sympathetic disease, has increased them. The linear method of extracting cataracts is one of these sources; although, happily, when we place in the balance the advantages and the evils of this operation, the former outweigh the latter. The operation of iridodesis is less fortunate, and raises doubts. The more recent operative procedure of *drainage of the eye* awakens still graver doubts concerning the propriety of its employment. Drainage of the eye consists of the insertion and retention of a gold wire through the tunics of the eyeball, with a view to causing a continuous escape of the fluid contents of the globe along the canal occupied by the wire. It was the hope of the advocates of this method of treatment that it would, on the one hand, prevent the re-accumulation of subretinal fluid, in cases of detachment of the retina, and on the other, keep within normal limits the intraocular pressure in glaucoma, and thereby become an effective therapeutical agent in both these affections. But the injury to the eyeball incident to this operation will seldom be tolerated, and notwithstanding the transitory relief obtained, an insidious inflammation of the uveal tract will be set up in the great majority of cases, with imminent danger of sympathetic disease. I have, in fact, learned without surprise, that where eyeballs have been drained by this process, it has often become necessary to enucleate them, on account of the sympathetic affections which they have induced.

SECTION IV.

PATHOGENY.

WE will first make a few *general remarks* on the pathogeny of the subject under discussion. The fact that a disease of any part of the body should be the cause of disease in a symmetrical member must in any event seem something extraordinary. Human pathology up to this day has revealed but few phenomena of this nature. Norris, however, in his paper on sympathetic affections of the eye, speaks of a few analogous occurrences in other regions; for example, one case by Mitchell, Morehouse, and Keen, in which, after a gunshot wound on the outer side of the thigh, complete anæsthesia was noticed on the corresponding side of the other thigh; and another by Annandale, in which, after a wound on one hand had healed with a painful cicatrix, a similar condition developed on the other.

Let us confine ourselves, however, to the eye, and at once inquire in what manner inflammation extends from one eye to the other. It would be an error to answer such a question in a general way. Entering therefore into details, we soon discover that the explanation is surrounded with difficulties of various degree, depending upon the locality of the inflammation. If we assume for example that the ophthalmoscope reveals an inflammation of the optic nerve and retina in the sympathetically affected eye, and that we are justified in assuming a similar inflammation in the injured eye (whose deeper structures we are usually unable to examine on account of entensive alterations in its anterior portion), we shall have no need of profound theories or the dragging in of obscure symptoms from other provinces of pathology, in order to understand what is going on.

In case pathological anatomy does not plainly inform us of any other way, we can assume in such a case, that the inflammatory process in the optic nerve of the offending eye propagates itself centripetally (toward the brain); the moment that the chiasma is reached, the optic nerve of the second eye is threatened. It is of no consequence whatever, in so far as concerns the explanation of the phenomenon, whether we are of those who claim a *total*, or of those who claim a *partial* crossing of the optic nerves at the chiasma; or whether we defend the view that *all* the fibres from one optic tract cross over at the chiasma to the optic nerve of the opposite side, or that a *part* of these fibres remaining on the *same* side, go to compose the optic nerve of the *same* side. For, in every case, the fibres of both nerves lie so close together at the chiasma, that it would be miraculous if the extension of an inflammatory process (particularly of the connective-tissue elements) were to confine itself, at the chiasma, to the fibres of *one* optic nerve, and carefully avoid the fibres of the second nerve which are so closely interwoven with those of the former. So far as concerns our present considerations, it is all one and the same, whether the process, after reaching the chiasma, advances or does not advance still further into the centre of the organ of vision, along the corresponding optic tract. But this much is certain: that, so soon as the fibres of the second optic nerve are attacked in the chiasma, the inflammatory process may extend not only toward the optic tract, but also toward the eye, and finally reach the terminal expanse of the optic nerve in the retina.

The appearance of typical irido-cyclitis in the eye originally affected, accompanied with the development of optic neuritis in the second eye, does not interfere with the explanation just given, for in such a case we take it for granted that neuritis (or neuro-retinitis) is simultaneously associated with the irido-cyclitis in the first eye. But how can we explain a sympathetic inflammation of the whole choroidal tract, and above all, sympathetic irido-cyclitis plastica, which many oculists consider the most important, if not the only significant symptom of the sympathetic affection?

We might imagine that under such circumstances, also, the inflammation was propagated *per contiguum*. Thus, irido-cyclitis may always be the primary affection in the eye first affected, while retinitis may be superadded to the original disease. The inflammatory process would then be simply transmitted along the tract of the optic nerves into the retina of the second eye, in which it could finally extend from the retina to the choroid. It is so common to see the choroid invaded by inflammation from the retina, that were a corresponding view permissible in the case of sympathetic affection of the uveal tract, all obscurities would be removed from the latter disease, and sympathetic inflammations could be regarded as simply transmitted continuously and *per contiguum* from the irritating eye through the chiasma.

Although the affection of the optic nerve, first in the one eye, and subsequently in the other, is still too little appreciated, it is nevertheless a fact that sympathetic irido-cyclitis does not originate by this agency. For, at the time when the premonitory symptoms of this latter affection appear, the retina is very rarely, if at all inflamed. Otherwise, why should not the most typical symptoms appear in the choroid proper, which lies throughout in immediate and extensive contact with the retina? In point of fact, it is the most anterior segment of the uveal tract (the ciliary body and the iris) which first suffers; that very portion

PATHOGENY. 109

which is covered by a merely theoretical part of the retina, the so-called pars ciliaris retinæ. As it thus appears that inflammation cannot be transmitted to the choroid of the opposite eye by the intermediation of the optic nerve and retina, we must either seek another path of communication, or else assume some remote and mysterious action.

There is, however, one possible path of direct communication between the two eyes. I refer to the vascular circle of Willis, lying in the region of the chiasma, at the base of the brain, corresponding to the sella turcica, and embracing the chiasma as well as the tuber cinereum and corpora mamillaria. Alterations in the choroidal vessels of one eye might be transmitted to the chief arterial trunk (the ophthalmic artery); from there into the internal carotid, and so to Willis's circle; thence along the anterior arch of this circle into the opposite ophthalmic artery, and so to the choroidal region of the second eye.

Cohnheim has already shown us what an important rôle is played in inflammatory processes, by alterations in the vascular walls; indeed in his opinion, "molecular alteration of the vascular walls," is the indispensable condition for inflammation. The only peculiarity with which we should meet in considering such a theory (even if all necessary assumptions were fulfilled) would be that the process in the second eye is never exhibited throughout the entire choroidal tract, but chiefly, or even exclusively, in its most anterior segment. Moreover, in the present state of our knowledge, we know nothing definite of any such direct transmission of inflammation along the vessels. By this, however, I do not mean to assert that the question of the participation of the vessels has yet been finally settled.

We have, therefore, nothing else to do than to keep to the nerves, under which term we of course mean simply the ciliary nerves. The short ciliary nerves contain motor, sensitive, and sympathetic fibres; and we shall assume that every short ciliary nerve is composed of fibres of each of these three varieties. The long ciliary nerves which arise directly from the nasociliaris have no motor fibres; of their sympathetic fibres we know nothing. Nevertheless, Stricker's experiments, which prove that hyperæmia is caused whenever we irritate the *sensitive* roots of the spinal cord (*i.e.*, that an irritation of the sensitive roots excites the nerves which dilate the vascular walls), would seem to show that the long ciliary nerves are made up in part of vascular nerves, which conduct irritation from the nerve-centre.

We are not inclined to acknowledge that the real motor nerves of the internal muscles of the eye, viz.: the corresponding fibres of the third pair, which supply the sphincter iridis and the ciliary muscle, as well as those fibres of the sympathetic which supply the dilator pupillæ, have anything to do with the transmission of sympathetic inflammation. There remains, therefore, for consideration only the sensitive fibres of the trigeminus, and the vascular nerves of the sympathetic. The question then arises, if the ciliary nerves are the only ones which act as conductors, does the capacity for transmission belong to each sort of fibres, or only to one, and to which? So far as concerns the motor nerves, I would say that we sometimes meet with simple paresis of accommodation, as the only symptom of sympathetic irritation (Pagenstecher, Mooren, Schiess-Gemuseus). This symptom, however, does not compel us to accept any action on the part of the motor roots. On the contrary, it can be explained in a very simple manner. The muscles of accommodation in both eyes contract synergically. If the contraction of one ciliary muscle becomes extremely painful on account of some morbid affection which has attacked it, contraction at once ceases, and with it also the contraction of its partner. But just so soon as the injured eye is enucleated, the ciliary muscle of the second eye at once resumes its function.

If it is the *sensitive* nerves which conduct the inflammation, we must assume that either some indefinable irritation, or an unknown molecular alteration, or a distinct inflammatory condition passes along the fibres into the brain, and reaches the central nerve-cells from which the fibres proceed; that this morbid process then "springs over" (or is perhaps transmitted by fibres) to the corresponding nerve-cells of the other side, and so, in turn advancing from the brain, reaches the terminal filaments of the sensitive nerves in the second eye. If the sympathetic fibres act as conductors, then the irritation must cross over to the other side, in the vaso-motor centre, *i.e.*, in the medulla oblongata, or, if we give any credence to Stricker's experiments, *beneath* the medulla oblongata.

It is relatively easy to assume some such state of things, for we thus safely avoid the dangers of "reflex" action. But, admitting that all this is proved, many difficulties still confront us, in our endeavor to explain the origin of inflammation in the sympathetically affected eye. The development of inflammation presupposes the fact that the irritation or inflammation of sensitive nerves can produce the most violent inflammation in the tissues to which they are distributed; or, relatively, that irritation of the sympathetic fibres which dilate the vessels, or paresis of the fibres which contract the vessels, not only causes an enlargement of the vessels (hyperæmia), but even true inflammation.

General pathology now busies itself but little with the influence which the nerves may exert upon inflammation, or denies it entirely. It is well worth observing that, from this point of view, so little attention, or even none at all, has been paid to sympathetic ophthalinia. Herpes zoster—a disease in which inflammation of the skin extends along the filaments of sensitive nerve-fibres—is the only well-known example of *the possible connection between an affection of the nerves and inflammation*, especially since the so-called neuro-paralytic inflammations—pneumonia after division of the par vagum, and keratitis after paralysis of the trigeminus—have been banished into the province of traumatic inflammation. And even as regards herpes zoster, Cohnheim thinks that we ought to wait for further and more careful anatomical or experimental investigations, before building conclusions of so great an amplitude upon a very few facts. On the other hand, no one has ever yet observed the development of a genuine inflammation as the outcome of that hyperæmia which depends upon division of the sympathetic nerve.

In considering sympathetic cyclitis, however, we must suppose some such direct influence of the ciliary nerves in the production of inflammation. In a clinical point of view, we have cases which afford such a hypothesis. In 1866 v. Graefe said: "It may be of interest to note the fact that in two cases of injury, in which I did not enucleate the wounded eye because it still retained some traces of vision, I was able, at the outbreak of the sympathetic affection, to prove that the second eye showed increased sensitiveness at a point, symmetrically to which a similar condition was present in the first eye during the whole period of observation." Bowman has also made one observation of the same nature.

Such exact symmetry as this is supposed to be extremely rare in ophthalmology, and even authors who have had at their command a large amount of material for the study of sympathetic ophthalmia, cite only the three cases of v. Graefe and Bowman. Despite this fact, I am, nevertheless, firmly convinced that this phenomenon is by no means rare. Still, it is always remarkably striking, no matter how often it may be observed. I have seen it in genuine iritis maligna, as well as in severe plastic iritis, in which the circumference of the iris had become bulged forward. It is also sometimes noticed in that sympathetic irritative condition which is usually regarded as ciliary neuralgia (page 63). If we carefully touch the region of the ciliary body of the sympathetically affected eye in these cases, we succeed in finding at some spot a pressure-point which is *chiefly or exclusively sensitive or painful* to the touch. If we then test the eye first affected, we are almost always sure to find an exactly corresponding spot over the ciliary region, which is chiefly or exclusively sensitive or painful. Although the originally affected eye frequently possesses but one painful spot, while the rest of the ciliary body remains quite insensible to the touch, or even to gentle pressure, so that, under these circumstances, it is sufficiently easy to discover the pressure-point in the eye first affected, we think it best to suggest that, in testing the sensibility of the ciliary body, we should begin in the eye affected secondarily. For the eye originally affected is sometimes so extremely sensitive to pain, that the attempt to discover if there be any especially painful spot in the ciliary region, *without knowing exactly where to seek for it*, is barbarous, to say nothing of the fact that it may be impossible of accomplishment. But the circumscribed pain from pressure, in an eye affected sympathetically, is not precisely the same sort of pain as that which is produced by pressure in an inflamed region of the body. It is much oftener discovered, on the contrary, as has already been suggested, even where we have nothing but a neuralgia of the corresponding ciliary nerves—a neuralgia which may disappear without passing into a state of inflammation.

If we reflect upon these facts, we can hardly do anything else than assume that the inflammatory irritation passes from the ciliary nerves of the one side to the corresponding ciliary nerves of the other, so that, finally, inflammation can be excited in the tissues to which these nerves are distributed. At present, however, in these cases, it is absolutely impossible for us to tell whether the inflammation is transmitted by the sensitive nerves, which are evidently affected, or by the sympathetic fibres. Herpes zoster seems to show an active participation on the part of the sensitive fibres; but we must not forget that, as sympathetic fibres are undeniably present in the ciliary nerves, we cannot, without further proof, deny the presence of the same sort of fibres in the sensitive nerve-trunks generally, as was demonstrated by Stricker's experiments, previously mentioned.

Having thus given a hasty and general glance at the subject, let us now see how the theory of the pathogeny of sympathetic inflammation has been built up in the course of time, upon the foundation of hypotheses, supported by clinical and pathological observations.

If Mackenzie was not the first oculist to recognize sympathetic ophthalmia, we may claim for him that he was the first author who published any papers that show deep insight into this terrible disease. As early as 1844 he had already developed various hypotheses concerning the pathogeny of this affection, which contain very nearly all that has been discovered in this province in the last forty years; while his works show that he had studied this obscure branch of ophthalmology much more carefully than is nowadays generally believed. For, in looking over his writings, we see at once that he had already considered the three paths along which sympathetic inflammation may possibly be transmitted: Firstly, through the vessels, by means of their anastomoses within the skull; secondly, along the ciliary nerves; and thirdly, through the retina and optic nerves. Nor do we now know much more about the manner of transmission than he did, for he says: "The vessels on the side of the injured eye, being in a state of congestion which may increase to inflammation, perhaps communicate a disposition to similar disease to the vessels on the opposite side, with which they anastomose inside the cranial cavity." "The ciliary nerves of the injured eye might be the paths along which the irritation is conveyed, through the mediation of the third and fifth pairs, to the brain, from which it is reflected along the corresponding nerves of the opposite side." And finally, speaking of the optic nerves, Mackenzie says: "It is extremely probable that the retina of the injured eye is in a state of inflammation which advances along the corresponding optic nerve to the chiasma. From there, the irritative condition to which the inflammation was due crosses over to the retina of the opposite eye, along its corresponding optic nerve."

Correct as this last view must appear, even in our days, Mackenzie undoubtedly erred in regarding the "union of the optic nerves" as the "*chief* medium" by which sympathetic inflammation is produced. For, although there is not the least doubt that sympathetic neuro-retinitis is often developed in the manner which Mackenzie pointed out, sympathetic inflammation of the uveal tract, as we have already seen, cannot be explained by the extension of an inflammation of the retina to the region concerned. So far back as 1849, Tavignot, as I learn from Mooren, regarded sympathetic iritis in the same light as if a sympathetic ciliary neuralgia were the primary affection, leading finally to hyperæmia and inflammation. V. Arlt also showed, at a later date, that conduction along the ciliary nerves was the more probable path: " We cannot decide, in the present state of our knowledge, whether, in such cases, the optic nerve (the neurilemma as far as the chiasma) or the trigeminus and sympathetic ciliary nerves are the intermediate agents, although a majority of facts speak in favor of the latter." Heinrich Müller (1858) was the first to awaken the attention of the ophthalmological world to the rôle that is played by the ciliary nerves. It is interesting also to note the fact that, from this time onward, the pathological views of sympathetic inflammation underwent very radical changes, although Müller's views differ so slightly from those held by Mackenzie. Müller, as well as Mackenzie, acknowledges that both the ciliary and optic nerves participate in transmitting the sympathetic irritation, but the former expresses himself in such a way that it seems as if he denied any such action on the part of the optic nerve. "Although I will grant that the ciliary nerves may often fan the fatal sympathy into flame, it is plain enough, at the same time, that I do not deny that sympathy (which assumes so many mysterious forms) cannot be transmitted by the optic nerve."

Although H. Müller followed in the general direction which had been indicated by his predecessors, his opinions seemed the more trustworthy because they were for the first time based on anatomical conditions. Among others of this sort, Müller found the ciliary nerves in a condition of partial atrophy, in an eye which had been enucleated on account of the premonitory symptoms of sympathetic ophthalmia. But, as the nerves had only lost their medulla, he thought that they might still have preserved "in a greater or less degree" their capacity for transmitting irritations toward the centre. "On the other hand," continues Müller, "the optic nerve, in many cases, is in such a condition of excessive atrophy, from the retina as far as the main trunk, that it could hardly have the power of transmitting an irritation, or any other process, from the eye." Nevertheless, we must emphasize the fact that Müller now spoils the effect of his last remark, by hastening to add that "we can hardly say, of certain fibres in the region of the lamina cribrosa, whether they are nervous or not." We must here carefully remark that Müller had not discovered any anatomical condition by which the propagation along the ciliary nerves could in any way be demonstrated; but that he simply based his conclusion upon the fact that the ciliary nerves are less liable than the optic nerves to degenerate into complete atrophy.

As years passed by, the opinion that sympathetic inflammation was transmitted by the ciliary nerves grew more and more fixed, while, during the same period, the theory of the participation of the optic nerves in the sympathetic process fell into oblivion. Pagenstecher (1862) was probably the first observer in Germany who wholly opposed the participation of the optic nerves, and referred the transmission exclusively to the ciliary nerves, chiefly to their "nutritive" sympathetic fibres. For many years thereafter the ciliary nerves were regarded as the sole conductors of irritation from one eye to the other. Nevertheless, a few men (among them Mooren) could not but notice many facts that tended to show some transmission along the optic nerve. In these exceptional cases only a secondary rôle was attributed to the optic nerves. Thus, in 1869, Mooren says that every sympathetic disturbance depends upon an irritation of the ciliary nerves, but that the trigeminus may affect the optic nerves in the following manner: the irritation transmitted from the trigeminus to the optic nerve of the eye first affected, might be carried along this optic nerve to the second eye; from the latter, in turn, it might extend from the optic nerve to the trigeminus, "so that the solution of transmitted irritative processes takes place in the ciliary ganglion."

But, beyond this obscure reflex action, it seemed to Mooren that a third factor was needed, in order to explain the origin of sympathetic affections: " one which fixes the relations of nutrition, secretion, and accommodation "—one which involves a co-operation of the sympathetic nerve, no matter whether the transmission is effected along the main branches, or directly along those sympathetic fibres which are said to accompany the optic nerve.

The first observer, of recent date, to claim that the optic nerve plays the chief rôle in the transmission of sympathetic ophthalmia is Alt, who bases his opinion on anatomical discoveries, which show a large percentage of alterations in the retina and optic nerve of the eye originally affected. We must not forget, however, that a large portion of these changes, such as the frequent occurrence of detachment of the retina, are nothing but the sequences of uveal diseases. We should mention, as an additional point of interest, that Alt also observed three cases of sympathetic neuro-retinitis. Finally, the same observer subscribes to the extraordinary opinion, that the whole nervous apparatus shares promiscuously in the transmission of sympathetic irritation to the second eye, and that the various types of the disease in question show only a difference of *degree*.

According to Mooren's theory, the nerves of *special sense* (that is to say, the optic nerves) would have to be additionally endowed with the capacity for *conducting irritation*. But if we assume that, at the time when the sympathetic symptoms appeared, there was no nervous connection between the foreign body and the optic nerve, and that *it would be impossible to prove any conduction through the optic nerve*, we should have to rely upon a different sort of (reflex) action between the ciliary and optic nerves, in order to explain certain sympathetic disturbances which are not of an inflammatory character. In the case already cited (page 67) of sympathetic contraction of the field of vision without any changes recognizable with the ophthalmoscope, Brecht expressed his opinion, on anatomical grounds, *that the optic nerves could not act as conductors*. Nor could he imagine any other path for the transmission of sympathy than through the ciliary nerves. Brecht also thought it quite probable that the foreign body might have excited inflammation in some of the ciliary nerves, which have the property of transmitting irritation toward the brain; that this inflammation extended step by step, and finally induced a hyperæmic condition in the medulla oblongata, with myelitis or some slight inflammatory process in the region of the vaso-motor centres. Subsequently, this inflammatory process caused paresis of the vascular walls, and hyperæmia of the retina in the second eye, which was the one at fault so far as concerned the disturbance of its function. Brecht based his argument on three experimental trials: first, those of Lewison on frogs (1869), from which the experimenter concluded that violent irritation of sensitive nerves paralyzes the reflex activity as well as those voluntary movements which are dependent on the medulla spinalis; secondly, on Leyden's opinion (1865) that the so-called reflex paralysis (paraplegia, paralysis of the sphincters), which is often observed after chronic affections of the bladder and other tedious diseases, may depend upon an inflammation of the sensitive nerves of the organ affected, which duly ascends into the spinal cord, and gives rise to a myelitis; and thirdly, on the experimental studies of Feinberg (1871), who observed paralysis of the bladder and paraplegia in a rabbit, a few days after cauterizing the ischiatic nerve, while at the post-mortem examination he discovered that the reflex action was due to a myelitis, the central stump of the cauterized ischiatic nerve being quite intact. This goes to show that a similar inflammation can gradually extend along the nerve. Moreover, it is to be regarded as an *experimental fact*, which confirms Leyden's discovery in man, that whenever he had diagnosticated, during life, a neuritis ascending into the spinal cord, he always found, after death, a corresponding myelitis at the place where the nerves entered, but no tokens whatever of an ascending neuritis.

We may here mention still another possible hypothesis. The well-known experiment of Golz, in which a frog's heart ceases to beat when one strikes a few rapid blows over the region of the belly, may be interpreted to mean that the centripetal sympathetic nerves of the viscera conduct a reflex irritation through the medulla oblongata to the vagus, which is the retarding nerve of the heart. Now, in the same way, we might agree with Brecht in supposing that the irritation due to the foreign body is simply transmitted, by reflex action, along the *sympathetic fibres of the ciliary nerves which lead to the brain (are there really any fibres of that sort?)* through the medulla oblongata to the ciliary nerves of the second eye, which lead from the brain, and that the latter then interfere with the function of the retina itself, just like any other retarding nerves. Leber also (1877) is of the opinion that, inasmuch as the reflex paralysis of motor nerves has been abundantly demonstrated, as well by clinical observations as by experiments on animals, the occurrence of a reflex paralysis " of sensitive nerves, especially of the optic nerve or retina," cannot at present be denied without further argument.

Those observers who defend reflex neuroses in the province of sympathetic affections, imagine, on the one hand, that the inflammatory irritation is undoubtedly conducted along the optic nerves, but that in the eye affected by sympathy the irritation crosses from the optic nerve to the ciliary nerves, by which the inflammation is first ushered in. Or, on the other hand, they assume that the sympathetic symptoms which reveal themselves on the part of the retina and optic nerve, are not produced in the second eye by direct conduction of the irritation from one optic nerve to the other, but by conduction along the ciliary nerves, and *from the latter to the optic nerve.* According to these views, therefore, the whole series of symptoms, such as sensitiveness to light, rapid weariness of the eyes during work, rhythmical indistinctness of the field of vision, periodical obscuration of vision, dread of light, sparks before the eyes, degenerating occasionally into excessive photophobia and photopsia, anæsthesia of the retina with concentric limitation of the field of vision, and finally typical retinitis (the latter separated from the other symptoms, at least by Leber, and regarded by him as the consequences of sympathetic iridochoroiditis)—all these symptoms, we say, are to be regarded simply as a series of reflex neuroses, the primary affection having its seat in the ciliary nerves.

The foregoing summary shows that we were right in designating our general views as relatively simple. But we will now go farther, and examine whether these relatively simple views will not satisfactorily explain all the phenomena of sympathetic ophthalmia without compelling us to enter upon the obscure province of reflex neuroses. When Mackenzie thought that there was very little doubt that the retina of the injured eye was in a state of inflammation, it seems as if he hit the mark precisely. Without being forced to assume some mysterious influence on the part of the ciliary nerves upon the optic nerves, it has now been proved that the injury itself is capable of exciting various inflammatory processes in the interior of the eye, and that they may (oftentimes, perhaps, from some definite lesion of the parts involved) rapidly attack the optic nerve. In this point of view, we find a very interesting fact in an insignificant remark of Brailey's, in his "Pathological Report for 1876." A boy, four years old, falls with a knife in his hand, and pierces the lower eyelid, and then the cornea, as well as a portion of the sclerotica right and left from the cornea. *Four days later* the eye is enucleated. The retina and choroid are both *in situ*. *The entrance of the optic nerve is swollen* and completely surrounded by a whitish opacity, near which lies a small capillary hemorrhage. The microscopic examination leaves no doubt of the swelling of the optic nerve. E. Williams reported at the International Congress in New York, in 1876, two recent cases in his own practice, in which the wounded and enucleated eye had been attacked, in the most surprising manner, by a very pronounced neuro-retinitis. In the first case (in which enucleation was performed a few weeks after the injury), Williams observed the most extensive swelling of the optic nerve that he ever had seen. Hirschberg also expresses astonishment over a similar case in the same year. In this case also, as in the one reported by Brailey, the eye was wounded by a knife-blade, although enucleation was not performed until nine months after the injury. The optic papilla was very much swollen, and surrounded by a well-developed wall, evidently due to hyperplasia of the inner granular layer, and the radiating fibres of the retina. Inasmuch as the development of the neuro-retinitis in the injured eye has been demonstrated by Brailey, at an early date after an injury, as well as at a later date by both E. Williams and Hirschberg, and since the frequent participation of the optic nerve in the inflammatory process in the injured eye has generally been confirmed by Alt, we have on the whole to take it for granted that the retina and optic nerve in the eye first affected are either irritated or inflamed by the wound itself, or by the morbid processes which follow the latter. It is, of course, hard to say wherein the "irritative condition" consists ; but it is a fact that the irritation can propagate itself to the second eye, or be produced in the second eye by inflammation in the first eye, as well as that the irritation can disappear after the removal of the original source of disturbance in the sympathetically affected eye.

Just in the same way as the obscurations of the

field of vision, as well as the diminution of central vision with concentric limitation of the field of vision, do not depend on diminished, but on increased irritability of the retina—not on anæsthesia, but on hyperæsthesia of the retina, so the sensitiveness to light, rapid weariness of the eye at work, photophobia, flashes of light and sparks before the eyes, are manifestations of irritation propagated from the one optic nerve to the other. The eye which has become over-irritated by the sympathetic process refuses periodically, or permanently, to react in various portions of its field of vision, to the irritation of an amount of light which would be plenteously sufficient for an eye in a state of normal excitability. And, further in this connection, we must remember that v. Graefe long since rightly referred to *hyperæsthesia* of the retina, that anæsthesia of the retina, with concentric limitation of the field of vision, which we observe in cases *where there can be no question of sympathetic irritation.*

Some one may ask how it is possible for such a connection to exist between the eyes, by means of the optic nerves, in those cases in which the optic nerve of the eye first affected is in a state of total atrophy. A cord of connective tissue cannot transmit such a sensorial irritation! Granted; but even if this is so, we cannot, in my opinion, assume with absolute certainty, in all those cases in which similar functional disturbances are observed, without any material foundation in the second eye, that *all* the fibres of the optic nerve of the first eye are atrophic. How could we decide, even with the microscope, that some minute fibres which still had the capacity of acting like nerve-elements, or axis-cylinders deprived of their medulla, might not still be present in the connective-tissue cord into which the optic nerve had become transformed? When Brecht, therefore, thinks it impossible that the optic nerves could have transmitted the sympathetic irritation in his case, and falls back on the ciliary nerves in order to support a theory of his own, he raises an unanswerable argument against himself, by saying that the eye first affected was *perfectly free from pain and irritation*. In other words, his supposition of an irritative condition of the ciliary nerves falls to the ground. We do not, however, mean to assert that the functional disturbances of the retina, which have been previously mentioned, do not depend upon alterations in the tissue concerned, even when the ophthalmoscopic image is negative. For we shall be compelled to assume some structural changes, even though they be coarse, when the irritation does not disappear after the source of irritation has been removed.. Thus, in Alfred Graefe's terrible case (page 65), in which the tormenting photopsies did not yield after enucleation of the injured eye, I cannot doubt that they originated in, and were kept up by, the products of *inflammation* which had already taken firm hold of the optic nerves. The microscopist, in these cases, gives us an important clue in this direction, when he finds *proliferated connective tissue* in the intraocular end of the optic nerve belonging to the enucleated eye. Such a proliferation of interstitial connective tissue in the tract of the optic nerve would gradually compress the bundle of nerve-fibres more and more closely, and finally give rise to mere mechanical irritation.

In previously speaking of *evident inflammation* of the optic nerve and retina of the second eye, we took occasion to emphasize the fact that there is no hinderance whatever to the transmission of such a process from one eye to the other. We had only to prove that such a neuro-retinitis was really present in the eye first affected. Indeed, I should like to believe that, when the retina and optic nerve of the first eye have been found intact after enucleation in a few cases of assumed sympathetic neuro-retinitis, this very fact alone takes away every point of support in favor of the sympathetic origin of the affection in question.

We now see why I so long ago emphasized the opinion that inflammatory affections of the nervous apparatus of the second eye really occur more frequently than observers have hitherto been inclined to admit, as well as that their presence is frequently hidden by the simultaneous appearance of irido-cyclitis; and, finally, that there is no necessity whatever for assuming that they simply indicate the extension of the inflammatory process from the choroid of the same eye. Nor should we forget, in speaking generally of the transmission of inflammation along the optic nerves, that this might also happen in case the optic nerve of the eye first affected were completely transformed into a thread of connective tissue. For, even in such a structure as this, the inflammation might creep onward to the chiasma, and then appear in the trunk of the second optic nerve in the shape of a dangerous peri-neuritis, embracing and crushing the fibres of the optic nerve by proliferation of connective tissue (a process which might finally reveal itself to the ophthalmoscope by partial or total atrophy of the optic papilla); or it might advance as far as the optic papilla, and there present itself to the eye of the observer under the form of optic neuritis. If we once hold fast to the fact that the optic nerve offers a very productive territory for the propagation of inflammation, we can then comprehend why optic neuritis may appear *in the second eye after enucleation of the first*, as in Colsmann's three cases previously cited (page 93). For, in these cases, the inflammation was either under way at the time when the operation was performed, and was only rapidly increased by the operation, or else the operation led to the neuritis by crushing the nerve during its division. Such a crushed condition of the nerve was indeed directly acknowledged by Mooren, in a case which he observed long before (1860) the cases cited by Colsmann. The patient began to complain of increasing dimness of vision, photopsia, and slight pressure in the forehead, a few weeks after the enucleation of the injured eye. Corrosive sublimate was exhibited internally, and a seton placed in the neck; but several months passed before the subjective symptoms disappeared entirely. The final history of the case showed that, two years later, atrophic alterations in the optic nerve (as confirmed by the ophthalmoscopic examination) had reduced the patient's vision so much that he felt fortunate in being able to read Jaeger's test-types No. 12, with difficulty. Who would not seek to explain such a case as this in the most simple way, by imagining that the operation gave rise to a peri-neuritis which extended to the second optic nerve, and produced partial atrophy?

We have, on the whole, no right at all to ask whether the sympathetic affection is transmitted along the optic nerves, or along the ciliary nerves; nor can we ask whether the transmission takes place along the one path more frequently than along the other. For the transmission may be effected in both ways. But by this, however, we are not to understand that one and the same morbid process can be transmitted, now along the one path, and now along the other. On the contrary, irritative and inflammatory conditions are

PATHOGENY.

transmitted from the optic nerve and retina, along the optic nerves; whilst those inflammatory processes which are chiefly observed in that portion of the eye which is nourished by the ciliary nerves, and especially in the uveal tract, are transmitted along the ciliary nerves. There is not the least doubt that the sympathetic inflammation may frequently be transmitted along both paths at once, or at short intervals, so that many symptoms in sympathetic affections of the uveal tract (amongst others, the functional disturbances) are not to be attributed to the inflammation of the uveal tract, but to a simultaneous inflammation of the retina and optic nerve.

This, of course, does not exclude the possibility of detachment of the retina, *appearing in connection with the irido-choroiditis*, involving the sympathetically affected eye, just as it may be observed in every irido-choroiditis. In the same way, when we see sympathetic neuro-retinitis in this same eye, the final detachment of the retina is not due to a sympathetic inflammation of the latter tissue, but to the process which is going on in the choroid.

Moreover, as any irritation of the stump of the nerve, external to the eye, can induce sympathetic neuro-retinitis, it is easy to see (if we once acknowledge that the ciliary nerves, or, in a wider sense, the branches of the trigeminus, can transmit the irritation) not only how cyclitis of the one eye can produ---

affections of the whole choroidal tract in the other, but also how the same morbid processes, which excite sympathetic affections in the *ciliary body* by irritating the ciliary nerves, can similarly become an irritating cause in other regions of the eye, as well as outside the eye, so soon as the filaments of the trigeminus, which are distributed to the regions concerned, are affected in an analogous manner. From all this we see that it is by no means extraordinary for irritation (incarceration), or inflammation of the iris, or of the choroid itself, or for the irritation caused by an artificial eye resting upon a stump, or finally, for the mere introduction of an artificial eye into the orbit after removal of the eye, to develop in the second eye about the same train of symptoms that we observe after a genuine cyclitis in the first eye. In the latter point of view (the influence of an artificial eye), Mooren was distinctly able to prove, in a case with great tenderness over the whole region of the stump of the optic nerve, how even a slight touch, upon the inner wall of the orbit, produced excessive pain—a fact which would go to demonstrate that the region to which the naso-ciliaris nerve is distributed was irritated by the sharp edges of the artificial eye. Moreover, a case of Snellen's, in which the sympathetic phenomena of irritation could at pleasure be excited and then dissipated, depending upon whether the artificial eye was inserted or again removed, shows how much these phenomena in the second eye may depend upon the irritation of the empty orbit by the glass shell.

Furthermore, we can see how enucleation itself, by crushing the ciliary nerves (and optic nerve) during their division, can become the starting-point of sympathetic inflammation, as well as how the curative reaction after a normal enucleation can excite the destructive disease in question by contracting the stump of the nerve in the cicatrix. In the same way it is easy to understand that, when the process in the first eye has once overstepped the rubicon, and is already advancing toward the chiasma along the extra-ocular tracts, enucleation cannot prevent its entrance into the interior of an eye which is still intact; and finally, that even when the cyclitis (or neuro-retinitis) in the first eye is entirely cured, the same process may subsequently appear in the second eye, and there continue its devastating course. The enemy had indeed wholly evacuated his first camping-ground, but at the same time he was already advancing rapidly upon the second eye.

Now, just as I have seen cyclitis appear in the second eye after complete recovery from the same disease in the other, or seen the second eye exhibit the most violent type of cyclitis despite the fact that the other eyeball was perfectly free from spontaneous pain, as well as insensible to the touch, it might not be at all impossible, after a normal recovery from enucleation, for some source of irritation to remain in the orbital or intracranial fibres of the nerve involved. I think that, in every case in which we have been obliged to ascribe the outbreak of sympathetic symptoms to the enucleation itself, or to the introduction of an artificial eye, we have, so far, observed, that the region in the bottom of the orbit which was occupied by the stump of the excised nerve, and its accompanying ciliary nerves, was sensitive to the touch, as well as that the conjunctiva lining the cavity was swollen, red, and painful. On the other hand, it would seem unjustifiable for us not to recognize the characteristic appearances of sympathetic irritation, as such, simply because up to this time we had never observed them in the absence of tenderness in the orbit, as well as at the stump of the nerve. I allude now to the following case:

March 25, 1878, I saw, for the first time, a farmer, aged forty-three, who had been wounded more than a year before, in the right eye, by the thrust of a cow's horn. A few days after the accident, violent pain was felt all over the corresponding side of the head. The injured eye was enucleated at a later date, but the pain did not cease. A year has passed since the enucleation, but the patient has never been free from exacerbating attacks of pain on the right side of his head. Still he does not seek advice so much for the pain, as because *his left eye is totally unfit for work.* He can use it so little, that it is only with the greatest difficulty that he can carry on his farm-work. He cannot read at all for more than a moment or two at a time. The eye looks normal externally and the ophthalmoscope does not help me to discover any internal alterations. The patient can read diamond type (Jaeger No. 1), and his field of vision is normal. The only definite anomaly which one can discover in the eye is that the power of accommodation is somewhat less than is usual at the patient's age. Despite, however, this nearly normal condition of the eye, the patient cannot work for any length of time, even with a convex glass to support his accommodation. We are therefore led involuntarily, in such a case as this, to assume the presence of a sympathetic neurosis. But when we examine the right orbit, we find that the cavity is lined with a conjunctiva which is neither red nor swollen, while neither in the bottom of the orbit, nor over the location of the stump of the optic nerve, can we discover any tenderness, nor even any special sensitiveness to pressure with a blunt probe. These various reasons had led several oculists to deny the possibility of any sympathetic affection in this case; but I do not regard it as entirely impossible. The irritative cause, even if the peripheral ends of the nerves show no distinct anomaly, may lie anywhere in the nerve-tracts; possibly even in the orbital portion of the optic nerve. In such a case, some remedy may yet be discovered by scientific investigation.

Another question now arises in considering the pathogeny of sympathetic affections: If we take it for granted that the nerves transmit the irritation, do we know anything more accurate regarding the *method* of transmission? We need not trouble ourselves because, in the present state of our knowledge, " it is impossible for us to know anything" about the molecular alterations which may be present in the nerves during the conduction of the irritation. But it is a more striking fact that we really know nothing more precise as regards the manner in which *inflammation* is transmitted. But even in this point of view we must distinguish between the ciliary nerves and the optic nerve.

Alt searched for alterations in the ciliary nerves in one hundred and ten cases in our province, but found only forty-three which offered any direct testimony. Thirty-four of these cases showed normal ciliary nerves. The remainder showed various lesions of the nerves in question, such as tearing, crushing (*without* histological alterations), incarceration in a cicatrix, fatty degeneration, atrophy, thickening of Schlemm's canal, and one case of calcareous degeneration in the same canal.

Goldzieher (1877) thought that he had unravelled the mystery, when he discovered in a given case (which in my opinion is very doubtful, so far as regards its genuinely sympathetic origin) such extensive alterations in the ciliary nerves of the enucleated eye as no other observer had ever before seen. The whole thickness of the choroid was filled with fresh inflammatory swelling and proliferation of cells; whilst the sheaths of the ciliary nerves were thickly infiltrated with round cells, and the inter-fibrillar tissues crowded with granules. Inflammatory nodules composed of round cells were also seen here and there compressing the trunks of the optic nerves. If such a condition as this were more generally observed, we should have at least some anatomical proof that the ciliary nerves are capable of propagating the inflammatory process within the eye, as has already been proved in the case of the optic nerves, even if we have, so far, been wholly unable to determine with exactitude the paths along which the inflammatory process is transmitted, outside the eye. But Goldzieher's discovery is very exceptional, and it cannot be denied that, in a vast majority of cases, the ciliary nerves of the eye which excites the sympathy show no alterations whatever. Goldzieher takes it for granted that the inflammatory alterations which he observed in the ciliary nerves are invariably present in such cases, and assumes, in correspondence with the experiments made on animals by Tiesler, Feinberg, Klemm, and Niedieck, that the inflammation in these nerves does not advance continuously, but by fits and starts, and that when it has reached the central organ it extends still farther in a similar manner. When the inflammation has finally crossed over to the nerve-tracts of the opposite side, it propagates itself in the same way, and so reaches in due season the network of nerves in the interior of the second eye, along which, in turn, the dangerous inflammation is conducted to the various membranes in correspondence with the distribution of the nerves concerned. So much for Goldzicher's opinion, to which we may reply that the theory of a wandering neuritis, as the anatomical cause of sympathetic inflammation, lacks at present any satisfactory basis, from the very fact that in almost every case the intraocular ciliary nerves are decidedly intact, to say nothing of the fact that no one has ever yet demonstrated such a wandering neuritis, nor proved how such an inflammation in a nerve (even were it demonstrated anatomically) could cause violent inflammation in a connective tissue.

Dark and complicated, therefore, as must seem the possible way in which inflammatory processes are transmitted along the *ciliary nerves,* the matter is relatively simple in the case of the *optic nerves,* for in the latter we have only to picture the transmission of an inflammation from nerve to nerve. Under such circumstances as these, the inflammation of the optic nerve, in the injured eye, is anatomically proved—in the eye affected sympathetically, it is directly proved with the ophthalmoscope; so that here, with the union of the optic nerves at the chiasma, we may calmly assume that we have to do with a connected or disconnected neuritis, passing from one nerve to the other through the chiasma.

Another important question for us to decide is this: How long does it take for the irritation which advances along the nerve-tracts to reach the second eye? This is about the same as to ask when the *sympathetic inflammation is liable to appear*. We may at once reply that we cannot fix the *latest* period at which the disease in question may make its appearance. If an eye is totally destroyed by an injury, the possibility of its reacting upon the second eye continues, not only so long as the eye is painful, but in case a foreign body has remained harmlessly in the eye (even at any region whatsoever), it may at *any indefinite future time* be followed by a reaction due to the presence of the foreign body (page 24). Or further, in an atrophic eye which, being utterly free from irritation, seems an extremely harmless neighbor, some unknown cause, or the development of a plate of bone in its interior, may give rise to renewed sensitiveness, and consequently develop a posthumous source of irritation (page 49). Finally, there can scarcely be any doubt that, in a *painless and unirritable* stump or eyeball, the seeds of sympathetic irritation can rest ungerminated for an indefinite period (pages 48 and 67). In point of fact, literature gives us the history of cases in which tens of years, even half a century, or longer periods, have elapsed between the original injury, or exciting cause, and the development of sympathetic ophthalmia.

It is much more important, however, for us to determine the *earliest* period at which the sympathetic affection may appear. In this point of view a proportionately long interval seems, in our opinion, to exist between the cause and the result. *A priori*, this interval cannot be measured. We have no precise starting-point from which to discover how long it takes for the morbid condition in the ciliary and optic nerves to be transmitted to the opposite side. So that, while the earliest appearance of *neuro-retinitis* in the injured eye has been precisely demonstrated, we do not know, so far as regards the ciliary nerves, how long a time must elapse before the ciliary nerves in the *primarily* affected eye are excited to the necessary irritative condition. We might even believe that sympathetic neuro-retinitis must necessarily be developed in a much shorter time than sympathetic cyclitis, because the path along which the cyclitis advances is much longer than in the case of the neuro-retinitis; nevertheless, we could by no means affirm that our experience corresponds to our expectation. Mackenzie stated that from one month to a month and a half separated the original from the induced affection, and I must emphasize the fact that, in my own experience, I know of no case in which I ever saw sympathetic ophthalmia appear sooner than in four weeks after the injury. I grant, indeed, that this period of four weeks might be somewhat shortened, in occasional cases, but I will not grant that the necessary period can be reduced *to a few days,* as is alleged to have been observed by several authors. There are, however, some observations after enucleation, which would seem to argue in favor of the possibility of the rapid development of the sympathy, although they deserve to be carefully examined. We saw (page 94) how Colsmann and Hugo Müller both observed one case each of *neuro-retinitis* in the uninjured eye *a few days* after enucleation of the other, and similar observations are at hand in respect to uveal inflammations (v. Graefe, Mooren, Schmidt, Pagenstecher, and Genth). But before we accuse enucleation of being the cause of the sympathy in these cases, we must prove that such an interval had not elapsed since the injury, as would have enabled the sympathetic inflammation to appear at that very same time, even if the enucleation had not been done, owing to the fact that the sympathetic irritation had long ago started on its path, and was just on the point of making its appearance in the other eye when the enucleation happened to be performed. When, in the case of the last two authors, the first symptoms of sympathetic iritis revealed themselves in the previously healthy eye nine days after the enucleation, we must remember that thirty-six days had already passed since the original injury—a period in which the outbreak of sympathetic ophthalmia cannot surprise us, for it could not, at that late period, have been restrained by an enucleation performed only nine days before. Schmidt's case is somewhat similar: sympathetic inflammation appears in four days after the enucleation; but here, also, nearly four weeks have elapsed since the injury. When enucleation is performed in the case of eyes which have *for a long time* been phthisical and painful (Colsmann and H. Müller), the sympathy which appears in a few days after enucleation can, with all the less certainty, be referred to the operation. So, if we have pure cases—*i.e.*, if one of two previously healthy eyes is seriously injured, sympathetic irritation will rarely appear before the fourth week; nor, when fairly under way, can it be restrained by enucleation.

The fact that a certain interval must elapse between the affection of the one eye and sympathy in the other is of great importance in establishing our diagnosis of a *sympathetic* disease. In order to make such a diagnosis in any given case, we must weigh well all that has previously been given in detail in these pages, under the sections of Etiology and Pathology. Furthermore, as we have already given a sufficient account of the general course and results of the more important types of sympathetic ophthalmia, especially as regards irritation, and the manifold forms of affections of the uveal tract, we can at this place dispense with any special remarks on the *prognosis* of the disease in question. And so much the more readily, as several points in this respect will be mentioned under the title of Therapeutics, to which we will now give our attention.

SECTION V.

THERAPEUTICS.

We finally turn our attention to the *therapeutics* of sympathetic ophthalmia, and instantly we hear the cry—I might almost say the battle-cry, "Enucleation." Scarcely twenty years have passed away since v. Graefe said: "I should never think it necessary to undertake the complete extirpation of an eye affected with traumatic irido-choroiditis, in order to ward off a sympathetic affection from the other eye, *and I only mention this operation because, as I hear, it is performed by some English oculists.*" Since then, thousands upon thousands of eyes have been sacrificed, and where is the oculist who feels wholly innocent of having operated under the philanthropical mantle of preventive enucleation, just for the sake of gaining some especially desirable specimen for his pathological collection?

Let us, however, enter calmly upon our discussion of this highly important subject. Before showing the *beneficial results* which enucleation may win for the patient, let us first inquire into the *harm* which it may cause. The most terrible result of enucleation (an operation which consists in shelling out the eyeball from its surrounding capsule of Tenon, sparing as much as possible of the conjunctiva of the globe, as well as of the external muscles of the eye) is—death! V. Graefe witnessed two deaths, when he enucleated during the period of *purulent panophthalmitis*, but none under any other circumstances. On the other hand, however, several fatal cases have been reported after enucleation of an eyeball *which was not affected with purulent panophthalmitis* (Mannhardt, Horner, Just, H. Pagenstecher, Verneuil, and Vignaux). The fatal cases reported by Horner, Pagenstecher, and Verneuil were due to meningitis, as was demonstrated at the post-mortem examinations, although in the first two cases there was no evident proof that the process had extended from the orbit; while in Verneuil's patient a phlegmonous inflammation of the orbit was proved to be the connecting link. I, also, once saw a fatal result after enucleation, in the case of an old woman whose right eye, after having undergone an iridectomy, continued painful, and had to be enucleated on account of absolute glaucoma. Profuse hemorrhage followed the operation, and death ensued in a few days. The orbit exhibited traces of suppuration, but there were no signs of meningitis. On the whole, there was no discoverable cause of death. There have undoubtedly been many more cases of death after enucleation than have ever appeared in print. For all that, we shall see how mere chance may play its rôle in this accident, from a case of my own, which will not easily be erased from my memory. An old woman had suffered for years with violent pain in a blind glaucomatous eye, which, with loss of sleep and appetite, had reduced her to a very feeble condition. At last she made up her mind to have the enucleation performed, and was received into the hospital. I postponed the operation for some reason or other, to the following day. But the operation was never performed, for on the morning of the day appointed, the patient was found dead in her bed. Had I operated on the day before, who is there who could not have said that the operation killed the patient? The autopsy in this case, as usually happens, revealed no cause for death.*

We are next to notice that the enucleation of eyes which are sacrificed in order to protect the second eye does not always progress without accidents, leaving

---

* As partially bearing on the question of chance, let us recall a case of our own, in which an iridectomy was appointed for a certain day, in a case of glaucoma. On the morning of the day appointed, the patient was found dead in her bed. Ought not the extremely few cases of reported death from *iridectomy* to be attributed to some other than the alleged cause?—Trs.

aside the very distant possibility of death. We may have extensive purulent inflammation of the orbital tissues without being able to discover any cause for such a course of events in the case itself, or in the operation; intense phlegmonous swelling, accompanied with violent pain, may be developed in the orbit and lids, compelling us to make an exit for the pus by extensive incisions into the orbital tissues and surrounding parts. At the same time, the general condition of the patient is weakened, and we can congratulate ourselves when the process confines itself to the orbit, so that all fear of its spreading into the cranial cavity is removed.

Again, enucleation always causes a local disfigurement, respecting the degree of which there may, however, be different opinions. Moreover, in so far as the eye removed had a certain size, and the operation was performed on a child, enucleation has considerable influence upon the configuration of the orbit concerned, as well as of the corresponding side of the face. There may of course be some discussion, in so far as regards the local disfigurement, as to which is the more comely, an empty orbit with sunken eyelids (which, however, every one will cover with a handkerchief or bandage), or a misshapen stump, which cannot easily or agreeably be kept constantly covered. To be sure, we shall hear in reply that the difference really consists in this: that an artificial eye, *fitted upon the stump*, satisfies the cosmetic demands more perfectly than when it is inserted into a vacant orbit. The artificial eye, a hollow glass-shell, with its concavity applied in corresponding size and curvature to the convex stump, deceives every one by the complete mobility which is imparted to it by the muscles still fixed to their normal attachments—a real eye, so true to nature as often to deceive even the specialist, if he does not look very carefully. It may, indeed, happen that the specialist himself mistakes the one for the other, the artificial for the natural, and the natural for the artificial eye. If the concave shell of the artificial eye is inserted into an orbit which has been deprived of its eye, the mobility of the former is not, as is generally supposed, completely abolished, although the motion which it really has is extremely slight. The operation of enucleation consists in removing the eyeball from Tenon's capsule. Now, the external muscles of the eye, in their course from their origin to their insertion on the globe, cross over to Tenon's capsule, and have to penetrate it in order to reach the sclerotica. But, at the very places where this penetration occurs, the tendons of the muscles become firmly united to the capsule. The investing membrane of the vacant orbit is chiefly composed of the conjunctiva of the eyeball, which now covers the capsule of Tenon; the latter in turn grasps the muscles firmly at the fissures through which they originally passed. Now, if the remaining eye moves, the corresponding muscles on the enucleated side also contract, so that some slight movements are still noticeable in the lining membrane of the empty orbit. These, then, are the motions which are partly transferred to the artificial eye, which is held firmly against Tenon's capsule by the pressure of the eyelids.

Although this tends to show that complete enucleation renders it impossible for us so well to satisfy the demands of good looks as in the case of a stump which still remains *in situ*, we must nevertheless remark that this circumstance is of but little importance in the particular series of cases with which we now have to deal. For, if we have the slightest dread of sympathetic irritation or inflammation in the well eye, we shall never dare to place an artificial eye upon a stump which is more or less painful; and even if we should by any means succeed in entirely freeing the eye from pain and irritation, we could never be sure of being able to apply the artificial eye directly upon the stump without the possibility of exciting sympathetic symptoms. Again, so long as the dangerous eye has a cornea, as may often happen, an artificial eye cannot well be worn; and besides, if the atrophic eyeball has not diminished considerably in size, the glass shell cannot be used.

Now, this cry of "mutilation" which has been raised by the opponents of too frequent enucleation, or of enucleation in general, cannot be accepted without a few words of explanation, for the early insertion of unbreakable artificial eyes may greatly compensate, in the case of a child, for the disadvantages of a vacant orbit, accompanied with a deformity of the face, or, more correctly speaking, for the inequality of development in one orbit and half of the face, in comparison with the other side. And, on the other hand, we must not forget that a minute stump will permit the very same aspect of things that we dread so much in the case of an entirely empty orbit. Thus, I have repeatedly seen so small a stump after blennorrhœa in the eyes of infants, that I was sure enucleation had been performed, and only after positive assurances to the contrary, was I able to discover in the bottom of the orbit a stump about as large as a pea, the convexity of which could not be seen, but only felt, beneath the enveloping conjunctiva. It is therefore a matter of no account whether a stump of such a size, or even somewhat larger, lies at the bottom of the orbit or not.

Death, cellular inflammation of the orbit, and a staring cavity (as well as other disadvantages of enucleation, such as excess of tears, and inversion of the lids, accompanied with irritation of the mucous membrane by the eye-lashes), have no direct relation to enucleation for *sympathetic ophthalmia*, but only to enucleation generally. The most important and most interesting question for us is whether enucleation in and by itself can do any harm; that is to say, can it endanger the other healthy eye by producing sympathetic irritation; or by increasing a *slight* form of sympathetic inflammation already present, to a more *violent,* or even the most violent form of all?

We have previously alluded to preventive enucleation in those cases in which the sympathetic affection appeared so quickly after the operation, that we could not but admit the possibility that the inflammation was already under way when the operation was performed. In such cases we can only say that the enucleation, at the most, hastened the sympathy, but did not really produce it. But the affair is quite different in those cases in which weeks or months elapse after enucleation, before the sympathetic symptoms appear. Thus, for example, enucleation was the starting-point of sympathetic neuro-retinitis in the two cases of Mooren's previously mentioned (pages 94, 132); it also caused a sympathetic "hyperæsthesia ciliaris" in a third case of Mooren's, in which the enucleation of an eye destroyed by a gunshot-wound had been long before performed. "The starting-point of the irritation in the present case must be sought for in the inflamed end of the optic nerve of the enucleated eye."

It seems to me, however, that we have much more important facts in those which tend to show that enu-

7*

cleation may increase those insignificant types of sympathetic affection which would never have greatly endangered the eye, to the most violent forms of sympathetic inflammation. Mooren (1869) enucleated an eye affected with cyclitis, because the premonitory symptoms of iritis serosa—" there were merely a few dots on the posterior wall of the cornea"—had appeared in the other eye. In the fifth week after the enucleation, Mooren for reasons unknown to us, made an iridectomy on the remaining eye, which was still free from pain. All went well for a time, but three weeks later—two months in all after the enucleation—a new and intense inflammation appeared, developed finally into a genuine plastic irido-cyclitis, and destroyed the eye.

Hasket Derby (1874) enucleated the eye of a young man with vision of $\frac{1}{40}$ normal, because three months after an injury the other showed simple iritis serosa (fine precipitates on the posterior wall of the cornea and slight dimness of vision). The deposits disappeared after the enucleation, and the eye, with normal vision, became again fit for work. But two months later irido-cyclitis appeared. Derby, suspecting irritation of the stump of the nerve in the region of the cicatrix, excised a quarter of an inch of the nerve, with its surrounding tissue. Improvement again followed, but did not last long. After several months, repeated attacks of iritis, combined with opacities in the vitreous, had reduced vision to $\frac{1}{10}$ normal. The final result must have been very sad.

Alt (1877) described the condition of an eye (in the case of a boy, aged nine years, injured seven years before by a needle) which was enucleated by Knapp for sympathetic iritis serosa. The behavior of the case after enucleation is interesting. The iritis serosa disappeared rapidly, but a plastic irido-choroiditis soon developed; vision sank to $\frac{1}{100}$, then increased to $\frac{1}{10}$. The termination of the case was unknown.

This transformation of simple iritis serosa into genuine irido-cyclitis after enucleation, is an extremely suspicious event. We have already drawn repeated attention (page 81) to the fact that iritis serosa, if not treated too heroically, does not seem to have any tendency to develop into the more severe forms of iritis, and I must confess that I cannot understand how Mooren (and others after him) can cite this case of his, as just quoted, as an argument against the opinion of v. Graefe and Donders, that iritis serosa never develops into iritis maligna under ordinary circumstances. Leaving entirely aside the fact that, in Mooren's case, an operation (iridectomy) was performed in the eye affected with iritis serosa, the ominous interval of two months between the enucleation and the violent inflammation, gives us a sufficiently distinct indication, not that the iritis serosa spontaneously increased to iritis maligna, but that the latter was caused by the enucleation (and would, perhaps, have appeared in precisely the same manner, even if the second eye, up to that time, had never been operated upon).

We see the same state of things in Derby's, Alt's, and in many other cases, in which enucleation in iritis serosa has been "fruitless"—that is to say, in which the second eye has been destroyed by plastic irido-cyclitis after enucleation of the first.

Samelsohn's case offers us a very instructive contrast to that of Derby, who, animated as he was with the best intentions, and guided by the opinions then prevalent, sacrificed an eye which still possessed vision, in order to save its partner, but lost both of the eyes; while, if he had not operated at all, both eyes might possibly have been saved. In Samelsohn's case, which is very similar to those just referred to, both eyes were really saved; not, however, by the skill of the surgeon, but by the persistent refusal of the relatives of the patient to have the proposed operation performed. We need hardly say, at this point, that we do not intend, in the slightest degree, to reproach the surgeons in question, but simply to utter our condemnation of those axioms according to which enucleation *must* be performed under such and such circumstances.

Here is Samelsohn's case in brief (compare Knapp's *Archives of Ophthalmology and Otology*, vol. v., p. 48): A boy of fourteen injures his left eye by a blow from the rebound of an elastic cord. Six weeks later fine dotted opacities appear on the posterior wall of the cornea, and, subsequently, a few delicate adhesions are noticed at the border of the pupil. The injured eye shortly before the last inflammatory attack could still read large letters (Jaeger, No. 23) with an excentric portion of the field of vision; finally, only fingers can with difficulty be counted. When the last attack in the left eye begins to decrease in intensity, the first symptoms of pericorneal injection, together with the characteristic opacities on Descemet's membrane, are noticed in the right eye. Enucleation of the left eye is now proposed, but energetically refused by the friends of the patient. Six weeks after the first appearance of the serous iritis, both eyes are not only free from inflammation, but from the least signs of irritation. The eye which had been affected by sympathy is perfectly normal. The injured eye has $\frac{1}{2}$ of normal vision, and shows only a slight contraction of the visual field.

In my opinion, there cannot be the least doubt that iritis serosa may become transformed into iritis maligna by the operation of enucleating the other eye. But, even as regards a slight attack of *iritis plastica*, enucleation cannot, under certain circumstances, be wholly acquitted of blame in furthering the transformation of the plastic into the malignant form of iritis. We must, however, make a separation between serous and plastic iritis. For, when we find a few adhesions in the second eye, before enucleation, while plastic irido-cyclitis develops itself afterward, we can say, with incomparably greater justification than if the case had been one of iritis serosa, that the posterior adhesions did indeed indicate the beginning of plastic irido-cyclitis, but that enucleation was simply unable to retard the process. We may be justified, moreover, in saying that the operation did not exercise any unfavorable influence. This is undeniably correct in some cases, but not in all. For we frequently observe cases in which the iritic process increases to irido-cyclitis at *such an interval after the enucleation*, that there can be no doubt that the plastic iritis, if left to itself, would have passed off as a mild attack, whereas the enucleation excited it to irido-cyclitis. We will here insert an appropriate case from Vignaux's rich experience: The eye causing the sympathy is blind, but entirely free from pain; the eye affected by sympathy is spontaneously painful, as well as painful to the touch over the ciliary region, and is affected with iritis accompanied with slight adhesions at the lower edge of the pupil. Vision is $\frac{1}{3}$ normal. With the help of atropia the iritis disappears after the enucleation. A month later, vision is fully $\frac{1}{2}$ normal. *Two months after the enucleation* a terrible inflammation appears in the eye, and, after persisting for ten months, leaves the organ in an incurable state of total blindness.

We have now uttered the paramount condemnatory opinion against enucleation—*i.e.*, that it may cause sympathetic inflammation in a previously healthy eye, as well as increase a mild inflammation to the most severe; or, more correctly speaking, that it may frustrate the permanent cure of a slight inflammation, by causing one of the most severe type. Hence, it is really only of secondary importance for us to add that, after the outbreak of a genuine iritis maligna, enucleation is not only of no benefit whatever, but that occasionally, when the sympathizing eye is extremely irritated, it really does harm; it even accelerates the disastrous process. Those cases of genuine iritis maligna which have recovered after enucleation, prove nothing at all in favor of the curative agency of enucleation, for no one will dare to say that in these extraordinarily exceptional cases, the process would not have proceeded in a possibly favorable manner even without enucleation, to say nothing of the fact that many such cases of perfect recovery rest upon an error in diagnosis: the case was not a genuine plastic irido-choroiditis.

Now that we have thus learned the disadvantages attached to enucleation, and the dangers which it may possibly have in store for the patient, it will be much easier for us to decide upon the importance of enucleation in the therapeutics of sympathetic affections of the eye.

The *fatal results* of enucleation do not trouble us much when we are deciding upon the operation, for the cases of subsequent death are altogether too rare. But, under certain circumstances, we still have some reserve in this respect. Almost all the German oculists hesitate to enucleate during the height of flagrant panophthalmitis, standing as they still do in dread of v. Graefe's two fatal cases (1863). This feeling goes so far, that a German operator even excused himself for having enucleated two panophthalmitic eyes *with the best results,* because he did not know at the time what v. Graefe had said on this point. Personally, I stand in awe of v. Graefe's advice never to operate if the panophthalmitis is distinctly pronounced. I have never enucleated an eye under such circumstances, and I doubt if I shall ever make up my mind to do so. The terrible apparition in v. Graefe's cases impresses me so deeply, that at the very sight of any eye in a state of panophthalmitis, and the thought of enucleating it, the dread of a fatal result is conjured up before me. By this, I do not mean to say that it is entirely justifiable for us to abstain from the operation, for the English oculists never pay any great attention to panophthalmitis when they desire to enucleate. Thus Critchett (of whom, as he himself laughingly said, the story goes that he cannot go to bed without having enucleated at least one eye during the day) told me that he had never seen an accident under the above circumstances. Vignaux also praises enucleation when thus performed; still he lost one case out of nineteen, although we must consider the great age (eighty-one) of the patient in this fatal case.

We do not mean in this place to treat of the general indications and contra-indications of enucleation, but only of enucleation as a therapeutical resource in sympathetic affections of the eye. Hence, we must justify ourselves for discussing enucleation in panophthalmitis. We have here brought up the subject, because, in our opinion, panophthalmitis cannot be *wholly* acquitted of the fault of producing sympathetic symptoms (although it is generally assumed to be innocent, on the ground that the acute purulent inflammation entirely destroys all the nerves in the interior of the eye). On the contrary, we are sure, that flagrant panophthalmitis may sometimes induce sympathetic inflammation, so that a few weeks after the outbreak of the original disease, and even at the time when it has by no means entirely disappeared, the premonitory symptoms of sympathy may reveal themselves in the other eye. Moreover, we mention enucleation in this place because when the panophthalmitis is excited by the presence of a foreign body remaining in the eye, we cannot expect a permanent condition of rest in the atrophic eyeball, even after the process has ended, but on the contrary, permanent or occasional spontaneous pain, or pain upon pressure, as well as the ever-threatening danger of sympathetic ophthalmia. So, if we venture to enucleate during the stage of panophthalmitis, we may not only put an end to the sufferings of the patient, produced by the acute inflammation, but secure him from the danger of sympathetic disease in the other eye for the rest of his life. But if any one is restrained from the enucleation of a panophthalmitic eye by the dread of a fatal result, the reasons which we have just suggested in favor of enucleation during this period, will not be urgent enough to overcome his fears. For the appearance of sympathetic ophthalmia *during* flagrant panophthalmitis, although observed by a few oculists, is so extremely rare as not to offer any general indications for the operation. In case, therefore, that the enucleation of the eye appears desirable as a precaution against sympathy in the future, we can wait until the panophthalmitis has gradually diminished under suitable treatment—in case we did not prefer to enucleate, or could not enucleate directly after the injury and *previously* to the appearance of the panophthalmitis.

Experience teaches us that when the irritation of the nerves has not yet extended to their extra-ocular branches, it is one of the rarest of exceptions for enucleation to lead to dangerous irritation in these latter filaments; and that whenever this does occur, the imperfect execution of the operation, or the crushing of the nerves during their division, is directly to blame in a considerable portion of the cases. We have, moreover, for the purpose of tabulation, only a very small number of cases in which we can say that the operator unwittingly caused the stump of the optic nerve concerned to become constringed in the cicatrix. From all these remarks we see that there is but slight probability of an intact second eye being endangered by enucleation of the first. And finally, so long as it has not been satisfactorily demonstrated, in any great number of cases, that enucleation increases a condition of simple irritation or mere disturbance of function to distinct inflammation, then, from this point of view also, enucleation is, on the whole, by no means to be dreaded.

To sum up our remarks, we have the following *indications and contra-indications for enucleation.*

If the second eye is still perfectly normal, oculists generally have not, up to this time, agreed upon the point whether preventive enucleation is admissible. My rule in such cases is as follows: if the patient is moderately intelligent, has good surroundings at his home, and can at any moment summon the counsel of a skilful oculist, preventive enucleation is not necessary. Some ophthalmologists claim that sympathetic inflammation can appear suddenly, and without any warning; but such is not my belief. The *intelligent* patient, warned of the threatening danger and notified to appear at once upon the slightest disturbance on the part of the sound eye, will hardly come to us with a pronounced irido-cyclitis, but at the first appearance of the slightest symptom of irritation. If, on the other hand, we have before us one of the lower classes, a patient defective in intelligence and in whom carelessness and mistrust of medical assistance are narrowly united; one whose remaining eye is liable to be overburdened with severe labor, and who cannot, even with the best intentions, get the advice of an oculist; then we may employ all our eloquence in favor of a preventive enucleation. For, notwithstanding our most earnest warnings, as well as all our representations that the patient will be totally blind for life if he neglects to report at the proper moment—despite all sorts of promises on the part of the patient that he will seek advice when the slightest irritation appears, we may never see such a patient again until vision shall have been irrevocably destroyed by a genuine attack of irido-cyclitis. Of what avail, then, to overwhelm the unfortunate patient with reproaches, to remind him of his promises, and even to fly into a passion, or to melt into pity, when he mildly says that he thought the eye would get well of itself, or that he sought help at the hands of some old woman!

The fact that the eye which is liable to cause sympathetic diseases at some future time *still possesses a certain amount of vision, never contra-indicates the performance of* PREVENTIVE *enucleation.* Those who resort to preventive enucleation on principle, or who regard it as a necessary duty to advise the enucleation of an eye in any special case, should never let themselves be led astray by the circumstance that the injured or irritated organ still possesses some remnant of vision. The enucleation of an eye which still possesses the faculty of sight, or one in which some degree of vision might *possibly* be restored at a later date, may be an unjustifiable deed in the general province of ophthalmology, but it can never serve as an argument in favor of abandoning *preventive* enucleation. For the removal of this eye assures the safety of the other, and no one should fear any subsequent objection to the operation. But frightful must be the silent accusation of one's conscience, when the patient in whom we regarded preventive enucleation as a necessity, but in whose case we were so weak as to be false to our convictions (simply because he still retained some vision in the injured eye), reappears before us with both eyes irretrievably lost. Read, for example, this case of Vignaux's: "A child about ten years old has received a blow on one of his eyes. Gayet is of the opinion that the eye should be enucleated, but abandons the operation because *the eye still possesses a certain amount of perception of light, and it is very hard to deprive such a young person of an eye which still offers some hopes for recovery of sight.* After a short time the child returns with the fully developed symptoms of sympathetic inflammation. The injured eye is enucleated; but it is too late; blindness becomes total." Gayet recalls this case to mind two years later, and says: " I shall regret this during the whole of my life." And I add, we hope that at the time when enucleation was finally performed, vision was really *wholly lost* in the injured eye, for if it were not, Gayet added to his previous error of abandoning preventive enucleation (one, by the way, in which, on account of the prevalent difference of opinions, he might find easy absolution) a second more grievous and much less excusable error, as shall soon be dilated upon more fully.

While discussing this point, I would like to add that I cannot see how Vignaux, while still depressed in mind by this case of Gayet's, could make such a remark as the following, one of the chief reasons against preventive enucleation : " Preventive enucleation is generally contra-indicated in case the second eye exhibits perfect organic and functional integrity, and *the originally injured eye still retains a certain amount of sight, or could obtain useful vision by operative interference at a later date.*"

If the general symptoms of sympathetic irritation are already present, enucleation *must* be performed *at once.* For, although cases have been known in which sympathetic irritation of the eye has lasted for years, and even decades, without really endangering vision, yet the physician cannot rely upon such a rare possibility in his own special case, in thinking over what remedy he shall employ. He must, on the contrary, regard the irritative symptoms as premonitory of the sympathetic inflammation, and, keeping in mind the danger that irido-cyclitis may be developed in a few weeks, even if no organic alterations are as yet present, he must decline all responsibility in the case, if enucleation is proposed to the patient, but refused. The oculist may act under such circumstances with energy and confidence; for, notwithstanding the few exceptional cases in which the inflammatory process is already under way, even here enucleation generally acts safely.

When the other eye is in a state of irritation, an eye which still possesses vision must be unhesitatingly sacrificed: success is too certain, and too much is at stake, for the oculist to hesitate. If, in such a case, he meets the rare misfortune of seeing the irritation become developed into inflammation despite the enucleation, he can say with confidence: "All is lost, but not my peace of mind." The surgeon *cannot* act differently, and such a tragic accident as just suggested is so rare that the vast majority of operators pass through life without meeting with such a lamentable experience.

If iritis serosa, and iritis serosa alone, is already present in the second eye, enucleation is, in my opinion, contra-indicated; and the enucleation, under these circumstances, of an eye which is not totally blind, is absolutely unjustifiable. I shall never again perform enucleation for sympathetic iritis serosa, for, as on the one hand this form of inflammation never shows any tendency to develop into irido-cyclitis, so, on the other, we have already offered proof of the deleterious influence of operative interference during the presence of this disease. In such cases, in all probability, enucleation does more harm than good to the second eye. Nor could I decide to enucleate in a case of *simple plastic iritis with a few adhesions, or even with adhesions entirely around the margin of the pupil*. We see a case like Vignaux's (page 158) in the one reported by Hirschberg (1874), in which enucleation was performed within a few hours after the outbreak of a simple plastic iritis in the second eye. The iritis proceeded favorably, but, about *three weeks after the enucleation*, a relapse occurred and the eye was finally lost. Even if Hirschberg is correct in assuming that the enucleation in this case was simply incapable of cutting short the irido-cyclitis which was already under way, the inexpediency of the operation would be evident. Under such circumstances enucleation cannot be of any advantage; it can only do harm. But we have already explained that plastic iritis is far from being synonymous with the primary stage of irido-cyclitis. For other reasons, however, a similar case of this sort will be mentioned farther on.

Inasmuch as enucleation undertaken during a violent inflammatory condition of the first eye is of no benefit in the presence of *sympathetic irido-cyclitis,* and may even rapidly increase the pernicious inflammation, it follows that, when we still desire to enucleate, we should wait until the inflammatory process in the eye which has been first affected begins to show some relative pause. There is no general indication for enucleation in cases of sympathetic irido-cyclitis. If, notwithstanding this, the eye is enucleated in this stage, the main idea can only be that where all is irredeemably lost, there is nothing more to lose. *Every one will admit that it is a crime in a case of pronounced sympathetic irido-cyclitis, to enucleate an eye which still possesses vision, or in which vision might at a later date be restored.* It ought to be absolutely impossible for any oculist to have the opportunity of congratulating himself, at the refusal of the proposed enucleation of an eye which still possesses vision while the other eye is affected with sympathetic irido-cyclitis; because the omission of enucleation under such circumstances should never be due to a lucky chance, but be dictated by the sagacity of the surgeon in charge of the case. Every one ought to know, and must know in such a case that enucleation cannot be of any avail. The oculist ought to know, even if there are several well-known cases in which irido-cyclitis has not led to total blindness after enucleation, that this favorable result was not obtained by the enucleation, but despite it. Moreover, he should be aware, on the other hand, that numerous cases have been reported, in which the eye causing sympathy has saved the patient from everlasting darkness, for the very reason that this eye still retained some useful vision after the eye affected by sympathy had become totally destroyed. V. Graefe said, after seeing two cases in which he *refused to enucleate* because the first eye was not totally blind : "I was extremely interested in these cases, by seeing perfect recovery from the sympathetic affection."

My creed in the question of enucleation runs briefly thus: It MAY *be performed as a preventive; it* MUST *be performed in the stage of irritation; it* CANNOT *be performed in iritis serosa and iritis plastica; it* CAN *be performed in irido-cyclitis plastica, provided the eye causing sympathy is totally blind, but not in a state of violent irritation.*

The most important point, so far as the general practitioner is concerned, is that he shall know the indications and contra-indications for enucleation. It is a matter of minor importance, whether, after having made a correct diagnosis, he can himself perform the operation, or feels obliged to refer the patient to a specialist for its performance. Still, I will in this place describe the details of the operation, as well as its after-treatment. Augustus Pritchard, of Bristol, England, was the first to enucleate a human eye for sympathetic ophthalmia (1851). The term "enucleation" owes currency in speech to v. Arlt, who proposed to use the term "enucleation of the eye" instead of "exenteration of the orbit;" that is to say, "the removal of the globe from Tenon's capsule," in contradistinction to the complete evacuation of the orbit, or the removal of the eyeball with all that lies behind it in the orbit. V. Arlt reserves the expression "extirpation" for the removal of some definite structure, such as a new-growth, from the orbit, with preservation of the eyeball. The shelling out of the eye from its envelope was first proposed by Bonnet (1841), and is performed in the following manner by v. Arlt.

Suppose that we intend to enucleate the left eye. The eyelids are kept apart by a stop-speculum, or, still better, by two lid-elevators in the hands of the assistant. In the latter case, by pushing the elevators along the lid, the assistant can separate the lids wherever the operator, for the time-being, requires the most room. The surgeon seizes the conjunctiva just over the insertion of the rectus externus muscle, with the forceps, divides it vertically with a pair of straight, blunt-pointed scissors, and then continues the incision in the conjunctiva half-way around the cornea and close to its upper edge, until he reaches the insertion of the internal rectus. He then returns to the original opening in the conjunctiva, and divides that membrane in a similar manner all around the lower margin of the cornea, but leaving a bridge of conjunctiva still standing at the inner side of the cornea, just over the insertion of the rectus internus. The next step consists in seizing the external rectus with the forceps, and dividing it completely; not, however, *between* the forceps and the insertion of the muscle on the sclerotica, but outside the forceps; or, more plainly still, between the forceps and the outer angle of the eyelids. In this way we have the stump of a muscle still attached to the eyeball, so that by seizing this with the forceps we can rotate the eyeball in any desired direction. One blade of the scissors is now directed upward beneath the tendon of the rectus superior, so that on closing the scissors the tendon of this muscle is completely divided from its attachment. After severing the rectus superior, the rectus inferior is treated in a similar manner. If we use a common stop-speculum, the assistant, having his hands free and possessing a sufficient degree of dexterity, can help the operator a great deal by taking up the tendons of the various muscles with the common strabismus-hook, and lifting them away from the sclerotica, so that it takes but an instant for the surgeon to pass the blade of the scissors between the sclerotica and the tendon, and to divide the latter completely. An operator of little skill, with an assistant of less skill, will of course help himself by taking up one muscle after another with the hook, before dividing them.

The three recti muscles (*the rectus internus yet stands*), with the conjunctiva which still covers them, have now been divided, or, more correctly speaking, the tendons of the muscles, as well as the conjunctiva, have been loosened from the eyeball. Now comes the most important step, the festal moment of the operation—the division of the optic nerve.

The optic nerve is inserted into the horizontal plane of the eye, but not precisely at its posterior pole; not at the posterior end of the antero-posterior axis of the eye, but a little toward the nasal side. In order, therefore, to pass deeply into the orbit with the scissors, the eye must be first turned toward the nose by means of the stump of the external rectus. But if the eye rolls at all on its antero-posterior axis, the insertion of the optic nerve no longer lies in the transverse axis of the eye, but approaches either the upper or the lower wall of the orbit. In order to strike directly across the optic nerve on introducing the scissors, we must be sure that the optic nerve remains in the transverse plane of the eye, which can only happen when we turn the eye *precisely* inward by seizing the stump of the external rectus. Hence, we must be sure to notice, when turning the eye inward, whether it rotates at all on its antero-posterior axis. If this should take place, we are to move the eye back again to its original position, and repeat the manœuvre nutil the correct position is reached. While the left hand is thus engaged, the right hand seizes a pair of strong, blunt-pointed scissors, curved on the flat, passes them (still closed) a short distance into the orbit along the horizontal plane of the eye, opens them, so far as is possible without resistance, pushes them forward, and closes them rapidly. A certain resistance on closing the scissors, a distinct, grating sound, extremely agreeable to the ear of the operator (for nothing is more disagreeable, in the operation of enucleation, than to miss the optic nerve), and the possibility of immediately lifting the globe out from between the eyelids, show that the operation has succeeded. But, if we have been so unlucky as to miss the optic nerve, we should not attempt to reach it by repeatedly opening and closing the scissors while in the cavity of the orbit. For the optic nerve now lies outside the scissors; it lies either above or below the latter. We should therefore remove the scissors entirely, once more carefully rotate the eyeball inward, and then repeat the manœuvre with the scissors.

When the optic nerve has been divided, and the eyeball drawn out from between the eyelids with the forceps, we take it in our left hand, divide the insertions of both oblique muscles, then the rectus internus, next the bridge of conjunctiva which still stands at the inner edge of the cornea, and the operation is completed; the eyeball, smooth and bare of all its attachments, with the optic nerve cut off close to the sclerotica, lies in our hand.

If the right eye is to be enucleated, we begin the operation over the insertion of the rectus internus, then divide the rectus superior and rectus inferior, leaving the bridge of conjunctiva standing at the outer side of the cornea. We should also remember that, on account of the insertion of the optic nerve on the nasal side of the antero-posterior axis of the eye, the nerve is found at a much less depth when we operate on the right eye, than is the case with the left.

The hemorrhage after the operation is generally slight. We may lay a couple of small plugs of charpie, cooled by contact with ice, into the cavity, apply charpie over the closed lids, and over all v. Graefe's compress-bandage (three or four turns of flannel), which is to be changed after twenty-four hours, and removed on the second day after the operation. In the course of recovery, the capsule of Tenon gradually becomes covered with conjunctiva, and in about a week we see at the bottom of the orbit nothing but a small suppurating and granulating surface, which soon cicatrizes completely.

The first thing of which we should be absolutely sure in operating for sympathetic ophthalmia *is to enucleate the right eye*. This may seem idle advice, and even a joke; but, whoever like myself has once stood shudderingly by, while the eye which still possessed vision was about to be enucleated instead of the blind eye, will not see a jest in these words of mine. The error is not inexplicable when we reflect that enucleation is frequently performed even when sympathetic cyclitis is already fully developed, so that there is really no obvious difference between the two eyes. Moreover, the operator is directing all his attention to the operation, and, being willingly led by the assistant, begins the operation on the eye to which the latter by mistake applies the speculum. The patient makes no protest—for he is under the influence of anæsthetics.

Anæsthetics have generally been resorted to in enucleation because the operation has been considered excruciatingly painful, especially during the division of the optic nerve, as well as of the ciliary nerves. I had always believed in this idea myself, and would scarcely have dared to enucleate without anæsthetics, had I not been compelled, in the case of a drunkard who really could not be chloroformed, to operate upon him in a conscious condition. I was not a little amazed when I found that the section of the various nerves was accompanied with no more acute expressions of pain on the part of the patient than during the first incisions in the conjunctiva. Since then I have repeatedly enucleated without anæsthetics, and have usually discovered, on questioning the patients after the operation, that the first incision (in the conjunctiva) was more painful than the division of the nerves. Mooren once went so far as to say that, "inasmuch as the operation is quickly performed, chloroform is used only when the patient expressly desires it;" and again: "besides this, I can operate much more easily if the patient is not chloroformed." At the time when I read these sentences, I was so firm in the belief that the division of the nerves was extremely painful, that I could not credit what Mooren had said. But recent experience of my own has shown me how true it all is.

Thus far for enucleation. The next question that comes up for our consideration is this: Inasmuch as the whole significance of the operation of enucleation depends upon the interruption which it causes in the conduction of irritation from the intra-ocular nerve-fibres to the extra-ocular branches, can we not gain precisely the same result by simply dividing the optic nerve (neurotomy)?

The history, in brief, of neurotomy for warding off or curing sympathetic ophthalmia is as follows: In 1857, v. Graefe said: "In order to decide whether the optic nerve takes an active part in the sympathetic processes of amaurosis, I have proposed in similar cases to substitute neurotomy for extirpation of the eye. Under precisely analogous circumstances we should, by adopting neurotomy, gain the advantage of preserving the eye." In 1865, Rheindorf reported a case of neurotomy performed for sympathetic neuro-retinitis, with scissors bent exceedingly on the flat, and rounded off at the points. Four days later the vision had increased by four numbers of Jaeger's test-type, and the recovery was permanent. The influence of the operation in this case could not be denied, for the excessive diminution of vision had persisted for months, during which period all treatment had been useless. The operated eye, at a later date, showed considerable injection of the anterior ciliary veins.

In 1866, v. Graefe returns to the question once more. Nine years previously he had proposed to divide the optic nerve, not as Mooren thinks, because " the celebrated suggester of this procedure meant also to divide the ciliary nerves," but because in these cases it seemed to him that the optic nerve served as a conductor. At this time, however, it is the section of the ciliary nerves which v. Graefe proposes, although he doubts the propriety of dividing *all* of them outside the eye, "on account of the necessarily extensive denudation, and especially on account of the simultaneous division of the vessels." On the other hand, in case of *circumscribed* sensibility of the ciliary nerves, we might divide such as were implicated, outside the eye, or perhaps better still, inside the eye, behind the flat portion of the ciliary body. Ed.

Meyer first performed such an *intra-ocular* division in 1866, and in 1867 and 1868 he reported this case, as well as several others in which enucleation would have been indicated as a preventive, or on account of irritation already present. A narrow knife is passed through the sclerotica into the vitreous, and a section six to eight lines long (depending upon the extent of the painful region), and parallel to the margin of the cornea, is completed by simple counter-puncture, and division of the overlying bridge of tissues. In 1868, Secondi also reported a case of radical cure of sympathetic neurosis by intra-ocular ciliary neurotomy. All the tunics of the eye were completely divided over a space of a centimetre or two in extent, between the insertion of the rectus externus, and that of the rectus superior. Lawrence also reported a similar case in 1868. Ed. Meyer afterward continued to operate in this same manner, and in 1873 speaks of twenty-two cases of which he has heard. He thinks that intra-ocular neurotomy is really indicated as a preventive, as well as in cases of actual sympathetic neurosis.

In considering the question of division of the ciliary nerves *outside* the eye, we are to distinguish between their division *with preservation of the optic nerve*, and *the simultaneous division of both the ciliary and optic nerves*. Snellen (1873) reports a successful division of some of the ciliary nerves behind the eye without doing any injury to the optic nerve.

The eye was totally blind, with excessive and circumscribed tenderness to pressure at the upper and outer margin of the cornea. V. Wecker (*Therapeutique Oculaire*) recommends this operative method for cases in which the injured eye possesses better vision than the one sympathetically affected whose vision is totally lost. In his opinion we ought not to enucleate under such circumstances, but we may divide the ciliary nerves which surround the trunk of the optic nerve. Nevertheless, it is not plain from v. Wecker's account that he ever really performed the operation.

The division of both ciliary and optic nerves behind the eyeball, as a general substitute for enucleation, was recommended by Boucheron in 1876, and subsequently by Schöler and Schweigger. Schöler thinks that this operation is entirely safe in all cases of threatening sympathetic ophthalmia, while Schweigger is of the opinion that enucleation is only beneficial as a preventive operation, and that, from this point of view, neurotomy is just as available as enucleation, which in his judgment has hitherto been opposed by the patient, on account of the dread "which the mutilation of one of man's noblest organs" must naturally arouse. Finally Hirschberg, although he once published a paper opposing neurotomy, subsequently convinced himself, in two cases, that it succeeded in relieving ciliary pain.

I would like in this place to make a few preliminary

remarks on neurotomies in general. It seems to me that it is only a complete extra-ocular division of all the ciliary nerves, as well as of the optic nerve itself, that can be relied upon in cases of sympathetic affection of the eye. It must be extremely doubtful whether intra-ocular neurotomy, *i.e.*, the partial slitting open of the eye as above described, ever permanently relieves the eye so treated, or offers absolute security against sympathetic irritation in the other, even if it is performed several times in succession or in one district after another. Spencer Watson (1874) cites a case which was operated upon by Ed. Meyer's method, in which the primary result was very satisfactory, but it was not permanent, and enucleation had to be performed at a later date. On the other hand, there is no operation by which we can *be sure* of dividing all the ciliary nerves *without doing any injury to the optic nerve*. As for myself, I can see no indications for such an operation; for, in the case suggested by v. Wecker, we must not only postpone enucleation, but every operation on the injured eye, for it may still be saved; whilst if this eye is blind, we must at the same time divide both ciliary and optic nerves for the purpose of terminating the irritation which they incessantly keep up.

Among the opinions of various operators, on the division of the ciliary and optic nerves, we may quote that of Mooren (1869): " I can hardly believe, in any case, that division of the ciliary nerves in the orbit can attain the purpose which its supporters claim for it; for, after fifty or sixty experimental operations for the division of various branches of the trigeminus, although I have usually seen a momentary and brilliant result, yet it has rarely been permanent. The desired effect disappeared as soon as the ends of the nerves reunited." V. Arlt also cites a case in the *Zeitschrift der Wiener Aerzte*, "in which he was sure that the ciliary nerves became reunited after once being divided." We have a perfect right to look at the subject from this point of view, for up to this time we have had no satisfactory assurance of the length of time during which the favorable result continues in cases of division of the nerves outside the eye. We can only assume that the ciliary nerves have been successfully divided when the cornea and ciliary body become totally insensible to the touch (or pressure) after the operation. Restoration of sensibility in either of these regions shows that the branches of these nerves had subsequently reunited. I will at this place report a case recently under my own observation, in which reunion did take place, and at a relatively early period.

A young man had been wounded in the left eye a short time before by a flying chip of wood. This eye now shows diminished tension; the ciliary body is sensitive to pressure. There is slight ciliary injection, the cornea is perfectly normal, the iris is dull in color, its periphery is bulged forward in knob-like processes, and the margin of the pupil is attached to a thick membrane which covers the pupil. Perception of light is entirely destroyed. The patient now comes for advice, complaining that for some time his right eye has been momentarily sensitive to light, and that he cannot use it for any close work. The objective examination of this eye shows that it is normal in every respect. As the left eye is liable at any time to excite sympathetic irritation, while the complaints which the patient now makes may be regarded as the commencement of this condition, optico-ciliary neurotomy (as Schöler proposes to call the operation which we are now discussing) is performed—October 30, 1880—instead of enucleation.

I open the conjunctiva over the tendon of the rectus externus, and extend the incision in an upward, and afterward in a downward curve, toward the insertions of the superior and inferior recti. I next take up the tendon of the rectus externus on the strabismus-hook, and carry the two ends of a catgut thread, No. 0 (armed with a needle at each end), through muscle and conjunctiva. I then divide the tendon, and hand the threads with the muscle and conjunctiva to the assistant, to draw down into the external angle of the eyelids. The next step consists in rotating the eyeball toward the nose, after which I penetrate, with scissors curved on the flat, into the cavity of the orbit, divide the optic nerve, and then alternately opening and closing the scissors, I denude the whole posterior surface of the globe as thoroughly as possible. The scissors are now laid aside. I then take a curved tenotome, push it into the orbit, and denude the posterior portion of the globe still more thoroughly, turning the eye again and again as far as possible toward the nose. The subsequent hemorrhage is comparatively slight. The rectus externus is now replaced and advanced by sutures; the two needles are passed through the conjunctiva (which was previously left standing near the margin of the cornea), then removed, and the ends of the sutures tied. Finally, a pressure-bandage is applied.

November 2, 1878, three days after the operation, the cornea has lost all its sensitiveness, and the ciliary body is insensible to pressure. The ciliary region is now considerably injected, and the patient complains of violent pain. The conjunctiva also is extremely congested and very sensitive to the touch. The sensibility of the entire cornea soon returns. The ciliary body continues insensible for a considerable length of time. On the last examination, however—December 10, 1878—the upper and outer portions of the ciliary body are distinctly painful to pressure. The eyeball is rather pale, deviates slightly outward, and is decidedly soft to the touch. The vague complaints about the uninjured eye continue. Finally, enucleation is performed by Prof. v. Jaeger. And what did we then discover? The stump of the optic nerve attached to the globe consisted of two parts. *The optic nerve had been wholly severed by the neurotomy, but the two ends had reunited; not indeed in perfect apposition, the two surfaces of the original incision being still in part plainly visible.*

The history of this case has also taught us the method by which the operation is performed. Schweigger divides the internal rectus in the middle of its insertion, instead of the external rectus, and reunites it after the operation with sutures, as previously described. After dividing the optic nerve, he rotates the posterior pole of the eye forward, by means of a small, sharp hook inserted into the sclerotica near the optic nerve, so that the insertion of the nerve is brought forward into view. In this way we can carefully denude the whole sclerotica, so that the ciliary nerves shall be divided without the shadow of a doubt. But are we sure that some branches do not reunite? If this should happen, it is not necessary for our purpose to take it for granted that the divided ends of the same nerve should always reunite with each other. The case which we have just cited does not testify absolutely in favor of the complete reliability of optico-ciliary neurotomy. Therefore the operation must be tested further, perhaps improved a great deal, before we can employ it with confidence as a perfect substitute for enucleation,

Meanwhile, we hope that no operator who puts full trust in it, and employs it as a preventive, in the belief that he thus insures the other eye from danger as thoroughly as he would do by enucleation, may ever be terribly undeceived by seeing a patient, in whom he has thus performed optico-ciliary neurotomy, reappear for advice at a later date, with all the symptoms of a genuine irido-cyclitis!

Among other operations proposed as substitutes for enucleation, we may next mention the production of purulent choroiditis by the early introduction of a thread into the threatening eye. It is said that, by passing a thread through all the tunics of the eye, and letting it remain until a slight serous swelling (chemosis) of the conjunctiva indicates that purulent choroiditis (panophthalmitis) has begun, the eye gradually shrivels and becomes insensible. Moreover, it is said that the danger of sympathetic irritation is thus entirely removed, owing to the fact that the purulent inflammation has more or less completely destroyed the ciliary nerves. V. Graefe refers, at three different periods (1860, 1863, and 1866), to this manner of producing artificial atrophy, which had, however, long before been resorted to for an entirely different purpose, in the case of hypertrophied eyeballs. Fener also has lately revived the same proposition. Just here, however, we have nothing to do with the influence of this procedure in diminishing the size of enlarged eyeballs, but only with its relations to enucleation. In spite of v. Graefe's recommendations, based, moreover, as far as we can see, on entirely theoretical grounds, we must emphasize the fact, which is easily evident from his own last words on this point, that he had never made any practical use of this method in cases of sympathetic ophthalmia. These are his remarks in 1866: "*It might, perhaps, be rational under certain circumstances,* especially after wounds or operations, when nothing more can be hoped for in the eye in question, to increase the diffuse purulent inflammation already present, by inserting a thread for two or three days. The patient suffers far less from the panophthalmitis (if soothed with cataplasms) than he would suffer from a subacute cyclitis, gains a less sensitive stump, which bears an artificial eye excellently, and finally is saved from the danger of transmission of irritation to the other eye."

But if this method really offers so great advantages, why had v. Graefe, up to that time, never resorted to it? It seems to me that he had some fear that it might act as a double-edged sword. For, saying nothing of the fact that even panophthalmitis, and the "less sensitive" stump, do not offer complete security against sympathy, the thread, although it might not increase the inflammation to genuine panophthalmitis, might cause cyclitis of a much more severe and dangerous type. Under such circumstances, this method might not only not remove the danger of sympathetic ophthalmia, but even favor the outbreak of this affection in the same way as, when a foreign body lies hidden in the eye, we cannot hope for a condition of permanent rest.

Is there any need of my giving anything more than a hint of the method proposed by Barton, which consisted in abscising the cornea, removing the lens, and subsequently applying poultices to the remnant of an eye in which a foreign body still lies encapsuled? Or shall I mention the proposition of Verneuil (1874), who, after unfavorable experience in four cases of enucleation, advises us to close the eyelids by uniting their edges (blepharoraphy), and illustrates the useful result of this method by two pertinent cases?

Barton tells us that, after abscising the whole anterior portion of the eyeball, and applying poultices for a few days, the foreign body, which has previously been lodged in the vitreous, is generally found lying somewhere in the conjunctival sac. This operation will, however, hardly take the place of enucleation, from the fact that it may possibly be followed by excessive secondary hemorrhage, as well as by violent and tedious panophthalmitis, so that the eyeball is gradually reduced to a minute stump. In Verneuil's cases, the irritation of the eye which led to sympathy on the part of the other, depended, as Laqueur has already remarked, on a lack of suitable protection.

Under similar exceptional circumstances, therefore, this operation may also be employed.

Iridectomy is the last operation to be mentioned. Are we to perform it on the eye which causes sympathy? Under one circumstance only: when the iris (the eye being otherwise unharmed) has become incarcerated in the peripheral wound in the cornea, after an injury or operation, as well as after spontaneous perforation of the cornea. In such cases we may have neuralgia of the eye first affected, or sympathetic inflammation of the second eye. Iridectomy is then of great benefit, for by this operation we can abscise the imprisoned bit of iris, as well as the crushed ciliary nerves, and succeed in saving both eyes from danger. But when the incarceration of the iris has already induced irido-cyclitis, or when the latter affection has originated from any cause whatever, iridectomy is of no avail, and cannot in any respect be advantageously resorted to as a substitute for enucleation.

When the sympathetic symptoms can be attributed to the crushing of the nerve during enucleation, or to secondary imprisonment of the stump of the nerve in the cicatrix, we may endeavor to remove the irritating cause by subsequent excision of the cicatrix. But even then we shall only gain permanent results under the same circumstances under which enucleation would originally have been beneficial. Thus, Hasket Derby reports a case of fully-developed irido-cyclitis which could not be cured by resection of the stump of the nerve (page 154); while, on the other hand, Mooren succeeded in permanently relieving the ciliary hyperæsthesia in his case (page 153) by some peculiar method (which may really have consisted in exsecting the stump of the nerve). In my own case (page 136) I proposed an operation to the patient, intending to dissect the optic nerve away from all its surrounding tissues as far back as the optic foramen, and then to abscise it. If the irritating cause were situated in the orbital portion of the nerve, we might, perhaps, succeed in relieving the tormenting pain from which the patient has suffered. Up to this time, however, my patient, to whom, of course, I could not guarantee perfect success, has not been able to make up his mind to consent to the operation.

We have now finished our discussion of the operations which may be practised upon the eye *originally* affected, but we have not yet exhausted our account of the operative therapeutics of sympathetic ophthalmia. We still have to inquire what operations, if any, are permissible *on the eye which has become affected by sympathy*. In these cases also it is important for us to separate the various forms and stages of sympathy. We cannot operate on the second eye so long as it is intact, or merely exhibits simple irritation, or slight functional disturbances.

Iritis serosa is the first affection of the uveal tract

that we are to consider. In general, this type of iritis will not need any heroic treatment, and we ought to act toward it with much greater reservation than in a case of the same disease which does not depend upon sympathetic irritation. For the sympathetic form is evidently dependent upon some irritation of the nerves, an irritation whose increase we dread so exceedingly that we always energetically oppose enucleation of the irritating eye, so long, at least, as the iritis serosa persists. When the common form of serous iritis continues for a long time, and will not yield to the usual remedies, we cannot do anything better than to perform iridectomy. But, just as we should not operate on an eye affected by sympathy so long as there seems to be no real danger from delay, so we should not be too hasty in performing an iridectomy in cases of sympathetic serous iritis. As v. Graefe said, in 1866: "I remember only two cases in which I felt obliged to perform paracentesis of the cornea, and once to perform iridectomy upward, in cases of obstinate iritis serosa. In all of these, however, the desired purpose was effected."

Simple plastic iritis with but few posterior adhesions of the pupillary margin, the intermediate portions of the iris reacting well to atropia, is to be placed on the same level with serous iritis, so far as the abstinence from operative treatment in sympathetic irritation is concerned.

We have, however, an exceptional state of affairs in cases of *total exclusion of the pupil by circular posterior adhesions*. Let us at this point recall our previous remarks on this subject (pages 76 and 80). The differential diagnosis between the condition in which the iris is bulged forward by the fluid of the posterior chamber on the one hand, or by the masses of exudation dependent on plastic irido-cyclitis on the other, lies chiefly, in our judgment, in the degree of hardness or softness of the eyeball, in comparison with the normal condition. If the fluids of the posterior chamber have bulged the iris forward, the eye will be doubtfully, or perhaps distinctly harder to the touch; if exudations have been at work, the eye will be decidedly soft. When the periphery of the iris is bulged forward in knob-like masses, the eyeball, however, being soft to the touch, the case is quite different from that in which, with similar appearances on the part of the iris, we can prove that the eye is harder than normal. This latter condition only is the one with which we are now concerned.

The literature at our command does not give a superfluity of advice for cases in which *sympathetic secondary glaucoma* is apprehended, or in the presence of symptoms which denote its approach. V. Wecker (1879) thinks that, "on account of the violent pain from which the patients often suffer in case of an attack of glaucoma after the development of complete posterior adhesions," we should confine ourselves exclusively to paracentesis of the cornea or sclerotomy; we should never think of touching the iris, or of performing iridectomy. "We shall not, as a rule, succeed," says he, "in loosening those fragments of the iris which adhere to the anterior capsule of the lens, and if we are so fortunate as to succeed in a few cases, the eye will be so much irritated by the contusion, that the momentary benefit which we seem to have won will be lost again by closure of the new pupil, and deterioration of the function of vision."

Unfortunately, I cannot assent to this view; for in cases of simple iritis, iridectomy is unnecessary, while in those in which the posterior surface of the iris has become adherent to the anterior capsule, the operation is hardly practicable. But in that condition of affairs which we are now discussing, there is no doubt that we can excise a piece of the iris with the effect of restoring the communication between the anterior and posterior chambers. By this means we may also successfully oppose the inflammatory attacks of secondary glaucoma, as well as of glaucoma itself, by removing the inducing cause. The following instructive clinical history may serve to throw light upon what we have just said.

A man about thirty-one years of age was seen at the Clinic April 30, 1876. On January 24, 1876, a cramp-iron had been projected against his left eye. The

9

patient suffered but little pain after the injury; the sight of the wounded eye was diminished, but he could still see pretty well. The eye was very sensitive to light, and a few days thereafter it began to redden. The "inflammation" passed off in a fortnight, but vision had at that time diminished still further. The patient kept at his work for another fortnight, but as it made the eye congested and painful, he applied a bandage over it and stopped work. Still a fortnight later, six weeks in all, after the injury, the right eye became affected, and was injected and painful. The inflammation continued with occasional exacerbations, so that vision was gradually reduced to its present amount.

The examination shows the following state of things in the left eye: A cicatrix, three or four millimetres in length, in which the iris has become incarcerated, lies in the sclerotica, at the outer edge of the cornea, just above its horizontal diameter. The iris, which is altered in color, and has partially lost its striated appearance, is tied down to the anterior capsule of the lens by numerous adhesions, whilst the pupil has been elongated toward the cicatrix in such a manner that it seems as if a regular iridectomy had been performed. The ophthalmoscope reveals the bright edge of the crystalline lens at the place where the iris is deficient. We know, therefore, that the lens was not dislocated by the injury. The vitreous is so full of floating opacities that we cannot get an image of the background of the eye. The whole ciliary region is slightly congested. The tension of the eye is not noticeably changed; *i.e.*, the eye is neither too soft nor too hard. Tactile exploration shows that the outer and upper portion of the ciliary region (not precisely in correspondence with the place where the iris is incarcerated) is sensitive to pressure. The sight of this eye has decreased to one-fourth, or, with a very weak concave glass, to one-third of the normal amount.

The right eye shows slight injection of the ciliary region. The pupil is completely excluded by posterior adhesions, and the periphery of the iris bulged forward, especially in the upper half of the iris, which is altered in color and appearance. The pupil is filled with a membrane which is thin and transparent at the centre, but thick at the circumference. The tension of the eye is perceptibly increased, but not to a high degree. *A sensitive spot, corresponding precisely in location to the one discovered in the left eye, is found by careful palpation.* Vision is reduced to one-seventh of the normal amount.

What are we to do? We cannot enucleate the injured eye, even did it possess only the slightest possible trace of vision. It is as clear as possible that we cannot enucleate one eye with *one-third* of normal vision in order to save the other, which at present has only *one-seventh* of normal vision, not even if we had any faith whatever in the efficacy of enucleation under such circumstances. On the other hand, I am restrained from operating on the eye sympathetically affected, by the dread which such an operation should always inspire.

The patient is sent to bed, receives a solution of atropia for his left eye (without, however, dilating the pupil), and a course of inunction is begun. A week later (May 6th), after three inunctions (not to these, but to the suitable regimen do I ascribe the benefit) the ciliary injection has disappeared from both eyes. The ciliary body in each eye is no longer sensitive to the touch. On the next morning, however, pain is felt in the right eye, increases all day long, and at night becomes very violent. May 8th.—The tension of the right eye (the one affected by sympathy) is noticeably increased, the lids are slightly swollen, ciliary injection is excessive, the cornea is slightly hazy, and the iris is bulged forward much more than at any previous time. Pain is also felt at the sensitive spot in the ciliary region (while the corresponding spot in the wounded eye is free from pain), and vision is reduced to counting fingers at one metre. In brief, the right eye exhibits all the symptoms of acute glaucoma. May 12th.—As vision has not increased, an iridectomy is made inward, a large piece of iris being excised. The incision heals, and the anterior chamber is restored. *The iris no longer bulges forward at its periphery, but lies in a plane.* The blood in the anterior chamber is soon absorbed, pain and sensitiveness of the ciliary body disappear, and tension becomes normal ; but the ciliary injection is still present (May 18th).

June 9th.—*Both eyes are perfectly free from irritation, and their tension is normal.* Right eye: The cornea is slightly cloudy near the cicatrix left after the incision, but is otherwise transparent. The newly formed pupil is partially covered with a membrane, which, however, permits light to enter the eye at its periphery. The iris lies in its normal position. Left eye: The floating opacities in the vitreous have decreased so much that the retinal vessels and optic papilla can be dimly seen by means of the ophthalmoscope.

The result of the case may be thus formulated in brief: *The injured left eye has one-half of normal vision; the sympathetically affected right eye, one-tenth of normal vision.*

"I always operate when the periphery of the iris bulges forward," as I said before in speaking of secondary glaucoma produced by sympathetic iritis. This operation consists, as is evident from the foregoing clinical case, in iridectomy, which has an undeniably beneficial effect. Sclerotomy, *i.e.*, the formation of a large wound in the sclerotica at the edge of the cornea, cannot be performed under the above circumstances (bulging of the iris), owing to the excessive protrusion of the periphery of the iris ; while, on the other hand, if it could be performed, it would not fulfil the indication of restoring the communication between the anterior and posterior chambers.

Secondary glaucoma after sympathetic iritis seems to me to be the only condition that allows of operative interference. For, as serous iritis, as well as plastic iritis, does not demand such treatment, in the same way we cannot operate during the height of irido-cyclitis, because by so doing we increase the morbid process which in turn rapidly leads to atrophy of the eye. The unfavorable results which I had obtained from iridectomy, when performed under such circumstances, led me over to the side of the large majority of oculists of the present day, who will not resort to any operation, not even to an iridectomy, in cases of plastic irido-cyclitis. When v. Graefe performed iridectomy "even in a simple condition of affairs," but like all other operators gained no beneficial results, he asked himself whether "the iridectomy might not have been performed at too late a date." Or whether "a broad excision of the iris toward the extreme periphery might not be of greater benefit, especially if we reflect, that when the iris has once begun to adhere to the anterior capsule of the lens, the adhesion advances rapidly toward the ciliary processes." In other words, v. Graefe inquired whether, if he made the incision in the sclerotica as in the cataract operation which goes by his name, the iris would not present itself more broadly, and in a more suitable position for being grasped by the forceps, so that a much larger piece might be excised.

V. Graefe's recommendation of such a method is based on the favorable result which he obtained in one case of this sort—the only one which he had opportunity of reporting up to that date. But many oculists have since discovered that v. Graefe's hopes were too sanguine. Mooren, for example (1869), expresses doubt whether even the *earliest and most successful* iridectomy can be of any avail at all in the malignant type of plastic irido-cyclitis, for in two cases in which he performed the operation at the very outbreak of the disease, and under relatively favorable circumstances, the result was fatal to vision.

Although a few cases of the favorable effects of one or repeated iridectomies in iritis maligna have since been reported (Hugo Müller, Grossmann, Pflüger), we must hold firm to the axiom, that only after the process has become entirely extinct (by no means sooner than a year after the outbreak of the sympathetic inflammation) can we decide whether an operation is to be undertaken or not. The condition of the eye after such a lapse of time is frequently a great deal more favorable than we should have deemed possible at the outbreak of the affection, and many an eye that a few weeks after the appearance of iritis maligna seemed to have fallen a prey to total atrophy, offers itself, at the end of a year, free from irritation, with proportionally fair tension, and prompt quantitative reaction to light, even when the pupil is blocked up; or, when the pupil is clear, or but slightly veiled, exhibits a surprising degree of vision. In the latter case, we should be well on our guard against operating with the intention of improving sight. In the former, on the contrary, we should not delay in our attempt to make a path for the rays of light to reach the retina. In such cases, however, we cannot expect any benefit from simple iridectomy, for the whole surface of the iris being adherent to the capsule of the lens, it is impossible to draw or tear away the iris with its adherent membranes. We can then only attain our object by simultaneously opening and removing the anterior capsule, giving rise at the same time to traumatic cataract. In other words, we must resort to "extraction of the lens, with simultaneous iridectomy and laceration of the false membranes." (V. Graefe.)

A narrow knife—*e.g.*, v. Graefe's cataract-knife—is entered at the upper and outer edge of the cornea, nearly on a level with the tangent of the highest point of its upper margin. It is next to be pushed through the iris, afterward *behind* the iris, and finally through the lens to a corresponding point of counter-puncture, so that the sclerotic coat is opened at the upper edge of the cornea by a linear incision ten millimetres in length. We then introduce the forceps in such a manner that one branch passes in front of the iris, the other behind it (really into the lens behind the anterior capsule, which is adherent to the iris), and try to draw the whole membranous mass between the lips of the incision, in order to excise it. In case the membranes will not follow the traction (we should not pull too forcibly), we must cut through the membranes, with a pair of fine scissors, in such a manner that a free triangular bit of membrane lies between the branches of the forceps, by means of which the bit can be removed from the eye. Then follows the evacuation of the lens, which, during this manipulation has already been broken up into small pieces. If the opening in the membranous iris closes again after the operation, or if irido-cyclitis attacks the eye which has lost its lens, we should (after opening the anterior chamber with v. Graefe's knife) simply divide the diaphragm by v. Wecker's forceps-scissors, one branch being passed through the iris and behind it, and the other lying in front between the cornea and the iris (iritomy). In the case described on pages 52 to 55, double iritomy enabled the eye affected by sympathy to see fingers at six feet (with proper cataract-glasses), while the other eye gained vision equal to one-eighth of the normal amount.*

---

* Pagenstecher (1881) is of the opinion that such an operation as is here described is a mistaken one, and that we can win much better results by making an iridectomy, and then removing the lens, together with its capsule, with a flat spoon.—TRS.

Little as we can expect from the operative treatment of sympathetic inflammation when this disease has once become well defined, and extremely probable as it is that more benefit can be obtained by refraining from operative interference, we have no reason to boast of the results of *medical treatment.* Serous iritis and simple plastic iritis (in and by themselves by no means greatly to be dreaded, as we have repeatedly urged) behave toward therapeutical measures like other types of iritis which are not of sympathetic origin. But therapeutics have no power over a genuine sympathetic irido-cyclitis. It is, indeed, extremely doubtful whether even the most energetic measures, whether mercurialization, or even acute mercurialization, in a case of the latter type, can save an eye, which, on the other hand, may recover without any employment of mercury whatsoever.

We may thus sum up our *therapeutical resources in cases of injuries of the eye* which may subsequently lead to sympathetic inflammation. If an eye is badly injured, a large portion of its contents evacuated, vision totally lost, and a foreign body undoubtedly present in its interior, it is best to enucleate at once, before the impending panophthalmitis makes its appearance. If the wound embraces a large extent of the eye, and we are sure that no foreign body remains behind (or, if the shape of the eye as well as a partial amount of vision has been preserved, even if it is probable that a foreign body is still lodged within the eye), we are not to be in too great haste to enucleate. We should rather put the patient to bed in a darkened room, and drop a solution of atropia into the eye at regular intervals. If we think it can still be of any avail, we should further add a compress-bandage; and lessen whatever pain is felt, by hypodermic injections of morphia. The application of iced compresses, as well as of leeches, notwithstanding their frequent employment, is really of doubtful benefit. It is only in the exceptional cases in which the patient cannot bear the pressure-bandage that we should resort to cold applications. We should, however, remove them the moment that they begin to feel disagreeable to the patient, and simply cover the eye gently with a bit of cotton cloth. If panophthalmitis ensues, we should leave the pressure-bandage on as long as the patient can bear it; afterward warm fomentations (thin compresses dipped in warm tea, or poultices of farinaseed or wheat-bread boiled) are indicated. We may also try Lelièvre's new poultice-papers, which are strongly recommended by Frommüller.

When the eye becomes purulent, excessively painful, and greatly swollen, we may attempt relief by opening it. But when the panophthalmitis begins to show signs of relapse, we should, as soon as possible, insist upon the renewed application of the compress-bandages.

If several weeks have passed since the injury (the patient having been kept perfectly quiet in the interval) and the panophthalmitis has diminished proportionately, we must examine the eye thoroughly to see whether it is now perfectly quiescent or not. If it should be quiescent, the patient may have our consent to resume his usual occupation, but should be warned most earnestly to take notice of the least return of pain in the injured eye, and to report for advice without a moment's loss of time. If, on the other hand, the eye is no longer spontaneously painful, but still continues sensitive to all slight external influences, as well as sensitive or painful to pressure, we should enucleate it at once. We should also enucleate the eye, even if it still possesses a slight amount of vision, provided that it cannot be securely guarded from noxious influences, or if we cannot rely upon the intelligence of the patient. But if the patient be thoroughly intelligent, we can point out to him the various symptoms and circumstances under which he should at once seek surgical advice.

As soon as the stage of *sympathetic irritation* has become pronounced, we should instantly enucleate, even if the injured eye still preserves vision. In *serous iritis*, as well as *plastic iritis with only a few adhesions*, we should never enucleate, but keep the patient under the most guarded regimen: rest in bed in a darkened room, regulation of the diet, together with care for easy evacuation of the bowels. Locally, we should resort to solutions of atropia. If the eye is painful, and the circumcorneal injection well pronounced (which conditions are, however, very rare in iritis serosa), we should try bloodletting at the temples, as well as poultices applied to the eye. Weeks, or even months later, when the iritis has wholly disappeared we may enucleate as a preventive of future evil, in case the exciting eye has not become wholly free from pain. If the inflammation has culminated in posterior adhesions, with bulging of the peripheral portions of the iris, and subsequent secondary glaucoma, we cannot rely upon the usual anti-glaucomatous remedies, such as eserin sulphate, pilocarpin muriate (in one per cent. solutions), but we must try to restore the communication between the two chambers by an iridectomy.

Genuine plastic irido-cyclitis demands, of course, the above-mentioned strictness of regimen, and the most abundant patience, as well on the part of the surgeon as of the sufferer. Bloodletting and atropia seem to do more harm than good in this type of the disease. We can best resort to repeated poultices, and (if necessary) to morphia injections. If the patient consents, we may try acute mercurialization, aiming to saturate the system with mercury in the shortest possible time. For this purpose, from six to ten grammes (3 iss.–3 iiss.) of gray mercurial ointment should be rubbed in daily, conjoining this treatment with the internal exhibition of calomel in one to two decigramme doses (grs. iss.–iij.) every two hours until salivation is produced. But inasmuch as irido-cyclitis rarely leads precipitately to unfortunate results, a common well-regulated course of inunction seems to me altogether more suitable. We ought to try this treatment in order to satisfy our consciences. But we should not expect too flattering results. If we carefully analyze the few reported cases of rapid and perfect cure effected by acute mercurialization after previous enucleation, we shall discover, without the shadow of a doubt, that the cases were not genuine irido-cyclitis, and that therefore this type of disease was not cured by mercury. Nor can we attribute any decidedly favorable influence to the enucleation. This operation, by the way, we can omit with a calm conscience under the circumstances here mentioned. The sympathetically affected eye may, if it has not become blind, be subjected to an operation at a future time.

We have now finished our account of the therapeutical measures which may be adopted in the severest forms of sympathetic ophthalmia (affections of the uveal tract), but we have yet to say something of the remedies which may be employed in the secondary or minor forms of this insidious affection. *Sympathetic retinitis or neuro-retinitis*, which ensues in company with inflammations of the uveal tract, cannot, on the whole, have any influence in inducing us to change our indications for operative interference, notwithstanding the few reported cases of sympathetic keratitis and scleritis have always been known to disappear after enucleation. This form of sympathy should be treated by rest, darkness, bloodletting, inunctions, and the iodide of potassium. Shall we enucleate if it is diagnosticated as being independent of any uveal affection? I am of the opinion that sympathetic neuro-retinitis is due to a similar morbid process in the opposite optic nerve and retina. Inasmuch, therefore, as the division of an inflamed nerve does not seem any too seductive to me, and as a relatively great number of these cases have been observed directly after enucleation (showing that the deleterious influence of the division, or of the cicatrix, upon the nerve can hardly be denied), I would not like to enucleate in a case of sympathetic neuro-retinitis, despite those favorable results which have been reported. Several cases of sympathetic retinitis were reported at the International Ophthalmological Congress, in New York (1876), by Alt, Derby, and Risley. Alt saw rapid improvement and recovery after enucleation in one of his three cases. But it seems to me that the sympathetic origin of these cases was not accurately demonstrated, for the optic nerve of the enucleated staphylomatous eye showed deep glaucomatous excavation and atrophy. Moreover, several observers besides myself have seen a sympathetic retinitis disappear spontaneously, under suitable circumstances.

INDEX.

A BSCISSION of cornea as a substitute for enucleation, 188
Accommodation, 16
" asthenopia of, 62
" impairment of, 68
Amblyopia, sympathetic, 97
Anæsthesia of retina, 66, 128
" sympathetic, 105
Anæsthetics during enucleation, 176, 177
Anatomy of ciliary nerves, 57, 58
" eye, 12
Anterior capsule, 13
" " incarceration of, in wound of eye, 42
" chamber, 13
Aqueous humor, 13
" " imprisonment of, behind iris, 78
" " normal means of escape of, 77
Artificial eyes, description and mode of adaptation of, 150, 151
" " as cause of sympathetic ophthalmia, 51, 70, 134, 136, 151
Atrophic choroiditis, 86, 92
Atrophy of ciliary nerves, 119
" eye, 28
" optic nerve, 119, 128
Atropia, 196, 203

B ANDAGE, compress, 175
Blepharoraphy as a substitute for enucleation, 188
Blepharospasm, 59
Blindness, total, from repetition of original injury, 9
Blood effused into retina, 88
Blows on eye, 25, 52

CANAL of Fontana, 16
" Schlemm, 16
Capsule of Tenon, 150
" lens, 13
" " shrivelling of, 42
Case of arrow-wound of eye (Mauthner), 21
" bit of iron encapsuled nine years in ciliary muscle (Bowen), 24
" " metal lodged seventeen years in optic nerve (Bowen), 25
" cataract operation producing sympathy (Mauthner), 39
" double operation for cataract (Knapp), 41
" enucleation after linear extraction of cataract (Mauthner), 39
" " for iritis serosa (Mooren), 154
" " " " (Derby), 154
" " " " (Knapp), 155
" " " " refused by patient (Samelsohn), 156
" " with unsuccessful result (Gayet), 165
" hæmorrhagic glaucoma (Pagenstecher), 100
" herpes zoster ophthalmicus producing sympathy (Jeffries), 44
" " " " " (Noyes), 44
" horse bite of eye producing sympathy (Mauthner), 29
" injury of ciliary body from bit of iron (Mauthner), 18
" " eye by a cramp-iron (Mauthner), 193
" " " from a bit of glass (Mauthner), 23
" iridodesis producing sympathy (A. Graefe), 35
" " " " (Steffan), 36
" leech bite of eye producing sympathy (Lebrun), 30
" lodgment of bit of metal in posterior chamber (Mauthner), 19
" neurotomy for sympathetic neuro-retinitis (Rheindorf), 178
" peri-neuritis of optic nerve producing sympathy (Mooren), 132
" persistent photopsies, despite enucleation (A. Graefe), 65
" poliosis arising from sympathy (Shenkl, Jacobi), 85
" primary lesion of optic nerve (Brailey), 126
" " neuro-retinitis (Williams), 126
" resection of stump of optic nerve after enucleation (Derby), 154
" reunion of nerves after neurotomy (Mauthner), 185
" sudden death associated with a proposed iridectomy (Translators), 148
" sympathetic choroido-retinitis (v. Graefe), 87
" " " " " 88
" " contraction of field of vision (Brecht), 67
" " ophthalmia after recovery from cyclitis (Mauthner), 52
" sympathetic ophthalmia from enucleation (Colsmann), 93
" " " " " (Mooren), 94

Case of sympathetic ophthalmia from enucleation (Müller), 94
" " . " " " (Vignaux), 158
" " " " gunshot wound (Cohn), 49
" wound of eye by a cow's horn, producing sympathy (Mauthner), 136
Cataract, cases of cyclitis after operations for, 39, 40, 41
" causes of original irritation in eyes treated for, 42
" depression (or reclination) of, 37
" division of, 37
" extraction of, 37
" flap operation for, 38
" lamellar, 32
" modified linear extraction of, 39
" operations for, causes of sympathetic ophthalmia, 31, 39, 103
" stationary central, 31
" sympathetic, 102
Cellular plate of ligamentum pectinatum iridis, 13
Chiasma, relations of optic nerves at, 106
Choroid, 15
" peculiar form of morbid patches in, 86
Choroidal sarcoma, 43
Choroiditis, 28
" atrophic, 86
" purulent, production of, as a substitute for enucleation, 186
Choroido-retinitis, 87, 88
Ciliary body, 15
" " detachment of, 42
" " diseases of, 17
" " foreign objects encapsuled in, 24
" " injuries of, 17, 23, 25, 26, 42, 65
" " " spontaneous cure of, 17, 18
" " " symptoms and anatomical changes caused by, 26
" ganglion, 58
" muscle, 15
" nerves, anatomy of, 57, 58
" " as conductors of sympathy, 110–120
" " atrophy of, 119
" " composition and functions of, 57, 58
" " reunion of, after neurotomy, 182
" neuralgia, 63
Circle of Willis, 109
Compress bandage, 175
Cornea, 13
" abscission of, as a substitute for enucleation, 188

Cornea, curvature of, impaired by iridodesis, 34
" paracentesis of, 193
" phlyctenulæ of, 59
" staphyloma of, 45, 46
Creed (Mauthner's) prescribing and limiting enucleation, 170
Crystalline lens, 12
Cyclitis, 26, 29
" acute, Mooren's definition of, 29
Cyclo-choroiditis as a cause of sympathetic ophthalmia, 46
Cysts of iris as a cause of irido-choroiditis, 43
Cysticerci, intra-ocular, 43

D EATH, after enucleation, 147, 160
Descemet's membrane, 13
" " deposits on, in serous iritis, 74
Detachment of retina, 28, 103
" ciliary body, 42
Diagnosis of sympathetic ophthalmia, 144
Diffused light, disturbance of vision by, 32
Diseases, sympathetic, list of, 56
" " relative severity of, 102
Drainage of eye, as a cause of sympathetic ophthalmia, 104
" " operation of, described, 103

E ARLIEST advent of sympathetic ophthalmia, 142, 143
Enucleation, accidents from, 147, 149, 152, 159
" after-treatment of, 175
" anæsthetics during, 176
" as a cause of death, 147, 160
" " " disfigurement, 149, 151
" " " sympathetic ophthalmia, 51, 62, 93, 132, 155
" creed indicating and contraindicating, 170
" crushing of optic nerve during, 162
" indications for and against, 163–170, 202–206
" in irido-cyclitis, 169
" " iritis maligna, 159
" " " plastica, 168
" " " serosa, 167
" " panophthalmitis, 160
" of left eye, 171
" " right eye, 175
" " wrong eye, 175

INDEX. 215

L ENS, crystalline, 12
" dislocation of, 34
" " as a cause of irido-choroiditis, 43
Ligament, suspensory, 12
" " laceration of, 34
Ligamentum pectinatum iridis, 13
Linear extraction of cataract, 39
" " " causing sympathetic ophthalmia, 41, 103

M EMBRANE of Descemet, 13
" " deposits on, in serous iritis, 74
Mercury in sympathetic ophthalmia, 196, 202, 205
Motion of artificial eye, 150
Motor fibres of ciliary nerves, 58

N EOPLASM attached to optic papilla, 22
Nerves, ciliary, anatomy of, 57, 110
" as conductors of sympathy, 110-120
" atrophy of, 119
Nerve, naso-ciliary, 57
" optic, 14
" " as conductor of sympathy, 110, 118, 120, 122
" " atrophy of, 119
" " excavation of intra-ocular end of, 97
" " hyperplasia of intra-ocular end of, 65
Neuralgia, ciliary, 63
Neuro-retinitis, sympathetic, 133
" " " treatment of, 207
Neurotomy, 177
" as a preventive of sympathetic ophthalmia, 186
" ciliary, 181
" extra-ocular, 180
" " general remarks on, 181
" " history of, 177
" " method of performing, 183, 184
" " reunion of nerves after, 182
" intra-ocular, 178, 181

O BLIQUE illumination of eye, 72
Occlusion of pupil, 76
Optico-ciliary neurotomy, 180
Optic nerve, 14
" " as conductor of sympathy, 65, 108, 121, 126, 130, 132

Optic nerve, atrophy of, 119, 128
" " excavation of intra-ocular extremity of, 95, 97
" " mode of crossing of, at chiasma, 106
" " sympathetic affections of, 92
" papilla, 14
Ora serrata, 14
Orbital cellulitis after enucleation, 149

PANOPHTHALMITIS, 47, 147, 203
" enucleation during, 160
Paracentesis of cornea, 193
Pars ciliaris retinæ, 109
Pathogeny of sympathetic ophthalmia, 105
Pathology of sympathetic ophthalmia, 56
Phlyctenulæ of cornea, scrofulous, 59
Photophobia, 59, 64, 128
Photopsia, 59, 64, 128
Phthisis of eye, 23, 100
Pigment spots in retina, 89, 92
Pilocarpin, muriate, 205
Plastic iritis, 75, 80, 157, 168, 191
Poliosis, sympathetic, 85
Posterior capsule, 12
" chamber, 14
Poultice-papers, Lelièvre's, 203
Preliminary remarks, 10
Pressure points, 84, 85, 113, 195
Prognosis of sympathetic ophthalmia, 145
Pupil, exclusion of, 76, 192
" occlusion of, 76

QUESTION of enucleation in sympathetic ophthalmia discussed, 146–177
Question of neurotomy in sympathetic ophthalmia discussed, 177–186

REFLEX action in conduction of sympathy, 112
Refracting media of eye, 12
Retina, 14
" anæsthesia of, 66, 128
" detachment of, 28, 103
" hæmorrhagic extravasations into, 50, 88
" hyperæsthesia of, 62, 66, 68, 128

Retina, irritation of, 62
" pars ciliaris of, 109
" pigment spots in, 89, 92
Retinal gliomata, 43
Retinitis, sympathetic, 87–92
" " treatment of, 207

SARCOMA of choroid, 43
Schlemm's canal, 16
Sclera. *See* Sclerotica.
Sclerotica, 13
" softening, or relaxation of, 45
" staphylomata of, 45, 46
Scleritis. *See* Sclerotitis.
Sclerotitis, 71, 207
Sclerotomy, 193, 197
Serous iritis, 71, 75, 80, 102, 167, 190
Sight, impairment of, without structural lesion, 66
" restored, of first eye, jeopardized by operation on other eye, 41
Staphyloma of cornea, 45
Staphylomata of sclerotica, 45, 46
Stricker's experiments, 110
Suspensory ligament, 12
Symmetrically painful points on eyeballs, 84, 85, 113, 195
Sympathetic anæsthesia, 105
" atrophy of optic nerve, 92
" cataract, 102
" choroiditis, 86
" choroido-retinitis, 88
" cicatrix, 105
" diseases of eye, varieties of, 56
" " " relative severity of, 80, 102
" fibres of ciliary nerves, 58
" glaucoma, 95–101
" " acute, 99
" " hæmorrhagic, 100
" " without inflammatory symptoms, 98
" iritis, from enucleation, 82, 93
" " maligna, 80
" " " mode of propagation of, 107
" " plastica, 76 *et seq.*
" " serosa, 71 *et seq.*
" irritation, 58–67
" " after foreign bodies in eye, 59

Sympathetic irritation, as affected by enucleation, 166
" " causes of, 61
" " condition of second eye in, 62
" " different forms of, 59, 60
" " of optic nerve, 64, 65
" " of retina, 63, 64, 66
" " removed by suppurative choroiditis, 180
" " with limitation of field of vision, 67
" " " impaired vision, 66
" " " phlyctenulæ, 59
" keratitis, 48, 70, 207
" opacities of vitreous humor, 102
" poliosis, 85
" retinitis pigmentosa, 92
" sclerotitis, 71, 207
" ophthalmia, after recovery from cyclitis without atrophy of eyeball, 52–55
" ophthalmia, diagnosis of, 144
" " definition of, 10
" " from artificial eye, 51, 70, 134, 151
" " " atrophy of optic nerve, 92
" " " bit of iron encapsuled nine years in ciliary muscle, 24
" " " bit of metal lodged seventeen years in optic nerve, 25
" " " cerebro-spinal meningitis, 44
" " " cyclo-choroiditis, 46
" " " cysticerci, 43
" " " detachment of retina, 47
" " " drainage of eyeball, 104
" " " enucleation of eye, 51, 62, 93, 132, 155
" " " glaucoma, 46
" " " glioma of retina, 43
" " " gonorrhœal ophthalmia, 50
" " " gunshot wounds, 49
" " " hæmorrhage into vitreous humor, 47
" " " herpes zoster ophthalmicus, 44
" " " horse bite, 29
" " " iridectomy, 42
" " " irido-cyclitis, 43
" " " iridodesis, 35, 103
" " " leech bite, 30
" " " mechanical injuries of ciliary body, 48
" " " operations for cataract, 31, 39, 103
" " " panophthalmitis, 47, 161

INDEX.

Sympathetic ophthalmia from prolapse of iris, 48
" " " sarcoma of choroid, 43
" " " syphilis, 44
" " " ulcers of cornea, 45
" " exsection of optic nerve in, 189, 154
" " iridectomy in, 189
" " medical treatment of, 202
" " relative frequency of, in the various cataract operations, 41
" " relative frequency of traumatic agencies producing, 26
" " time of appearance of, 141
" " without cyclitis, 51
" " " disease of uveal tract, 51
" " " injury of ciliary body, 50
Sympathy, means and methods of transmission of, 105, 132, 138
" transmitted by the ciliary nerves, 110, 111, 115, 117, 118, 120, 139
" transmitted by the circle of Willis, 109
" " by the optic nerves, 106, 117, 119, 127, 140
" " by reflex action, 120, 125

TENON'S capsule, 150
Tension of eye, definition of, 21
Therapeutics of sympathetic ophthalmia, 146
Transmission of sympathy by ciliary nerves, 110–120
" " circle of Willis, 109
" " optic nerves, 65, 108, 121, 126, 130, 132
Traumatic complications, in diseases of ciliary body, 17
Tunics of eye, 12

ULCERATIVE process permitting prolapse of iris or of ciliary body, 45
Uveal tract, 15
" " acute purulent diseases of, 46, 47
" " idiopathic affections of, 43
" " mechanical irritation of, 43

VISION, impairment of, without anatomical changes, 66
" restored, of first eye, endangered by operation on second, 41
Vitreous humor, 12
" " filamentous opacities of, 102
" " molecular opacities of, 88

WILLIS, circle of, 109
Wounds of ciliary body, 17
" eye, arrow, 21
" " gunshot, 49, 50 68

YELLOW spot, 15

ZONULA of Zinn, 12
" " laceration of, 34

THE END.

www.ingramcontent.com/pod-product-compliance
Lightning Source LLC
Chambersburg PA
CBHW031814220426
43662CB00007B/638